聚
Communal Forums

姚仁喜 | 大元建筑作品 30×30
KRIS YAO　ARTECH SELECTED WORKS

30×30

| KRIS YAO ARTECH | 姚仁喜 大元 建筑工场 |

姚仁喜 著

辽宁科学技术出版社

This three-volume monograph contains 30 selected projects spanning 30 years of KRIS YAO | ARTECH's works. This monograph is a testimony of our efforts and devotion in constructing contemporary architecture – shaping sense of place, developing space drama, and elevating cultural context. The monograph is organized in the following three sections:

Cultural Scenes exhibits the projects that emphasize celebration of cultural and historical roots,
Communal Forums focuses on the spaces that inspire creativity in our pulsating society, and,
Social Sanctuaries presents works that conveys inherent tranquility for reflection and contemplation.

Summer, 2015

本套书精选姚仁喜 | 大元建筑工场成立 30 年来多种类型的 30 件作品，分别以三大系列综合呈现，记录我们对于构筑当代形式、经营场所精神、发挥空间戏剧与提升文化意涵的努力。

"艺"空间经堂入奥，以坚实的构筑实体，彰显历史文化的人文氛围；
"聚"空间着重于人之聚合，关照并提供常民生活的舞台；
"思"空间借由静谧建筑的力量，构筑安定、沉净的心灵场所。

2015 夏

Kris Yao has demonstrated his humanity with the development of physical elements
that hold true to human emotion and conditions,
cultural and historical context, and a sense of scale and place.
His architecture has a poetic nature,
using his native eastern aesthetic and spirituality
with a sense of natural light,
interplay of surfaces and forms and executing all with a high level of innovation and professionalism.

Commendation by the American Institute of Architects (AIA)
at the 2014 Honorary Fellowship Awards Ceremony

姚仁喜建筑师利用建筑元素的创作，
具体掌握了人类情感与生命状态、
文化语境与历史涵构、场所精神与人性尺度，
充分彰显了他的人文精神。
他以来自东方的美学与心灵的涵养，
以素材、造型与自然光的交互辉映，
加上高度的创意与专业的执行力，
建筑因而盈溢诗意。

美国建筑师协会 2014 年于芝加哥颁发姚仁喜建筑师荣誉院士之颂词

CONTENTS 目录

FOREWORD 序文 **8**
A TREATISE AS REFINED AS THE SPRING WINDS AND
AS PURE AS AUTUMN WATER BY *ZHOU RONG*
春风大雅，秋水文章 / 周榕

16 ARCHITECTURE AS THEATRICAL STAGE
BY *KRIS YAO*
建筑是舞台 / 姚仁喜

CONTINENTAL ENGINEERING CORPORATION HEADQUARTERS 18
汉德大楼

32 QUANTA RESEARCH & DEVELOPMENT CENTER
广达研发中心

KELTI CENTER 52
克缇大楼

72 HUA NAN BANK HEADQUARTERS
华南银行总部

CHINA STEEL CORPORATION HEADQUARTERS 88
中钢集团总部

106 HUMBLE HOUSE
寒舍艾丽酒店

HSINCHU HIGH SPEED RAIL STATION 122
新竹高速铁路站

154 HSINCHU BUS TERMINAL
新竹转运站

YUAN ZE UNIVERSITY LIBRARY 166
元智大学图书馆

176 LUODONG GOVERNMENT CENTER
罗东行政中心

CHRONOLOGICAL LIST OF SELECTED PROJECTS 190
精选作品年表

196 COLLABORATORS
合作团队

FOREWORD

A TREATISE AS REFINED AS THE SPRING WINDS AND AS PURE AS AUTUMN WATER

Zhou Rong, Ph.D.

Associate Professor, Tsinghua University School of Architecture, Beijing

The pinnacle of Chinese culture is elegance and refinement, the essence of five millennia of civilization and Chinese spirit, an art form imbued with beauty, yet cultured in its romanticism. Chinese refinement is grand and profound, and has continued to reinvent itself throughout the ages. Its everlasting qualities reveals how it has blended the new and old throughout the ages, surviving wars and turmoil, and witnessing the rises and falls, all without yielding. Moreover, the ability of Chinese culture to continue along the path of refinement and elegance is not limited to its language, but instead extends to encompass all forms of creation. Therefore, although architecture is thought to be but a lesser art, there exists within it wondrous grace and refinement. That is the reason why true architects strive industriously to create works that would rise above mere craftsmanship, while inheriting the legacy of the refined. They are charged with the responsibility of carrying on this precious cultural heritage.

As the standard-bearer of his generation, Kris Yao has deeply influenced Taiwan's contemporary cultural landscape over the past thirty years, and his impact has since spread far and wide across the Taiwan Strait. Yao's architecture has become important benchmarks in both modern Chinese architecture and in the continued formation of modern Chinese culture. Therefore, to understand Kris Yao, it would be unwise to limit oneself to the confines of Taiwanese architecture; one should examine his work within the modern transformation of Chinese civilization in order to fully demonstrate the underlying historical value of his works.

Beginning in the late Qing Dynasty, facing the turmoil of the invasion of western civilization, Chinese civilization was forced to initiate a difficult transformation to modernity. A century of change brought vicissitudes and hardship. The political divide of 1949 splintered Chinese civilization as the two sides of the Taiwan Strait went their separate ways in regards to modernization. Those who crossed over to Taiwan were able to preserve the hierarchy of traditional Chinese culture. This enabled traditional "refined culture", which is an amalgamation of "elite values", to flourish in the Taiwanese society.

The orientation of the grassroots and the elite exacerbated the underlying differences in cultural perspectives between Mainland China and Taiwan over the past six decades. Taiwanese society held on to the core values of refinement and kept its cultural values from collapsing under the attack of the foreign, using refinement to adapt to the times, linking the new with the old while maintaining elegance.

The modern refinement of the traditional Chinese culture bred in Taiwanese society is reflected by the literature of Pai Hsien-yung and Chiang Hsun, the films of Hou Hsiao-hsien and Ang Lee, the dances of Lin Hwai-min, the plays of Stan Lai, and the architecture of Kris Yao. They have demonstrated alternative visions and possible paths for the thoughtful modernization of Chinese civilization. For their peers in mainland China, these works, though not necessarily grand, can be viewed as alternatives for the modern transition of Chinese culture.

Yao's contributions to cultural refinement is epitomized by the taming of the "red-haired wild horse" from the West – in this case, modern Western architecture. Architectural forms and vocabulary are transformed in accordance to the core values of traditional Chinese culture and re-organized into Yao's unique refined and cultured spaces. Under Yao, the refinement of Chinese culture is rendered by employing the concepts of "harmony" from Confucianism, "transcendence" from Taoism, and "stillness" from Buddhism. Through this, Yao is able to absorb and utilize modern Western architectural forms while lessening the bluntness of their brute force.

Confucian harmony influences the foundation of Yao's spatial designs. Harmony eliminates contradictions, avoiding the forced regimentation of hyper-efficient, complex spaces and unified logical forms that modern architecture tends to impose on the user. Under his precise and rational orders, Yao is able to moderate redundancy to create free, comfortable spaces. His balanced approach is free of the extremism in Western architecture, exuding instead a relaxed, human warmth. Arrogance is not present in Yao's buildings, nor are there complicated or enigmatic concepts too difficult for the layperson to understand.

Taoist transcendence is expressed as freedom in Yao's designs. This type of freedom is different from the aggressive freedom in contemporary Western architecture, which asserts individuality and destroys in order to create. Yao's freedom is passive, just "a dip in a pool of clear green water" without ripples. Traces of his restrained freedom is evident in the subtle variant-script bay window in the Lanyang Museum, the almost indiscernible shifts between symmetry and asymmetry of the overhanging wall on the west side of the Water Moon Monastery, and the ever-changing patterns in the wooden lattice windows in the Wuzhen Theater. Even in his bolder works, such as the free flowing Palace Museum Southern Branch in Chiayi and the New Taipei Art Museum, Yao still maintains a certain delicate balance between freedom and order. Transcendence is like flowing water, blissful, but not abusive; free, but not indulgent. Kris Yao's architecture transcends craftsmanship into the realm of art, from art to Tao, balancing yin and yang creating lively, exuberant refinement and grace at the behest of rational order.

Buddhist stillness as a temperament is unconsciously revealed in Yao's designs due to his personal adherence to Buddhist philosophy. Not to mention the Buddhist temples such as Luminary Buddhist Center and the Water Moon Monastery, even in bustling creations such as the Hsinchu High Speed Rail Station and the Wuzhen Theater, Yao exhibits a high degree of restraint in his expression of substance. His intentions are presented subtly, filled with compassion for the illusions of a fleeting life, paired with a hint of reluctant to build for what is merely ephemeral. Kris Yao's realm of refinement is in between the Risshu and the Zen schools of Buddhism, or rather, a taste of Zen in Risshu. A fleeting moment of youth and beauty creates eternity within the transient, just like the lone image of the Kelti Center in the Xinyi District; it may appears very close, but is actually far away; it seems to move but is still and quiet; it looks real but is illusory. A fashionable veil cannot hide the original face of the immovable; even though many turns of tides have come and go, a gap as thin as the cicada's wings is still hard to cross.

Qing Dynasty poet Deng Shiru once wrote: "like the spring wind, a gracious person can hold many things / like the clear water in autumn, a piece of writing won't be contaminated by dust"; and as mentioned in the *Doctrine of the Mean*, "all living creatures can grow together without causing each other harm, and paths can run parallel without interfering". The spring wind encompasses all living creatures and all parallel paths in harmony and with great tolerance. Kris Yao's architecture integrates modern Eastern and Western architecture and fuses Confucianism, Taoism and Buddhism; he has abandoned the complicated for the simple and omitted the contrived for candor.

序 文

春风大雅,秋水文章

周榕 博士

清华大学建筑学院副教授

雅乃文华,五千年文明精神之萃聚,钟灵毓秀,蕴借风流。中华之雅,博大精深,生生不息,与古为新,纵历乱离颠沛、兴替轮回而弦歌未绝。华夏文明之雅续,远逾文字,更广寓于诸般造物。故建筑虽为小道,亦有大雅存焉。是以建筑师孜矻营造,非止匠艺,兼挑雅脉,高致深情,寄诸土木。

作为台湾现代建筑的一代旗手,姚仁喜先生三十年来,不仅用自己的建筑作品深刻改变了台湾当代的文化景观,其影响更跨越海峡,成为中国现代建筑乃至中华现代文明创造的重要标杆。因此,认识姚仁喜,仅仅局限在"台湾建筑"的小格局内观察显有不足,应须将其放置于中华文明现代转型期的大参照系中进行审视,方能充分彰显其工作的历史价值。

晚清以降,面对西方现代文明入侵的巨大"灾变",中华文明被迫开始艰难的现代化转型。百年鼎革,沧桑困苦,神州裂变,同向殊途:1949之后,一水相隔的中华两岸步入了截然不同的现代化轨道。"衣冠南渡"的台湾,相对地保留了中华传统文明的层级结构,因此令凝聚了"精英价值观"的"雅文化"在台湾社会得到了较为系统的存续。

草根与精英两分的不同文化取向选择,在很大程度上造就了大陆和台湾六十余年来文明景观的底层差异。台湾社会持守雅文化的价值内核,得以在"俗"与"洋"两大现代浪潮的夹击下不至引发精英文化体系的全面崩陷,从而令这块割据的中华文明领地展现出一种"雅致现代化"的发展可能性。以雅化时,接古通今,是谓"雅续"。

中华文明传统的"现代雅续",在台湾文化界通过白先勇、蒋勋的文字,侯孝贤、李安的电影,以及林怀民的舞蹈、赖声川的戏剧、姚仁喜的建筑等外化形式绽放流溢,显影出中华文明走向现代化未来的另类愿景与可能通路。这簇彼岸文化精英的优雅创造,对雅育断代的大陆同侪和晚学来说都不啻是一种替代性探索实验。他们的作品虽非鸿篇巨制,但其对中华文明现代转型的多样化路标意义却已足够深远。

姚仁喜对文明雅续的贡献,集中体现为他在设计中对现代建筑这匹来自西方的"红鬃烈马"所进行的形式"雅驯"——即按照中华传统雅文化的核心价值准则，去筛取和改造现代建筑的形式语汇，并将其重新组织成温文尔雅的姚氏空间语言体系。在姚仁喜笔下，中华文化之雅具体呈现为向儒中取"和"、道中取"逸"、释中取"寂"。借此，姚仁喜对西方现代建筑形式既利其用，亦化其戾。

儒家之"和"，确立了姚仁喜雅境设计的温度与基调——散淡自洽、消弭矛盾的"协和"，避免了现代建筑高效复合的功能空间及统一严密的逻辑形式对使用者过强的规定性，通过适度的冗余处理在谨严的理性秩序中仍营造出一派空间自在；执两用中、不极不偏的"中和"，消解了西方当代建筑中常见的强调极端化表现和戏剧化冲突的内在紧张感，从而让物质环境释放出松弛的人情暖意；内敛自守、知柔处弱的"谦和"，使姚仁喜的建筑从无俯视众生的傲慢自矜，也没有普通人难以理解的艰深繁复，而是放低智识姿态，空间揖让成礼、设计收放合度。一团"和"气，方能涵纳万物，氤氲成雅。

道家之"逸"，在姚仁喜的设计中表达为一种"弱自由"。这种弱自由迥异于西方当代建筑中以张扬个性为目标、以冲决破坏为特征的强劲狂放的积极自由，而是"点破一泓澄绿"但却"从心所欲不逾矩"的低烈度的消极自由。从兰阳博物馆点到即止的破体飘窗，到农禅寺正立面西侧悬挑墙面在对称与非对称之间进行的若有若无的扰动，以及乌镇大剧院将千变万化的图案纳入规则控制的木棂花窗中，都可以发现这种低调自由的浅痕淡影。即便是逸兴遄飞如嘉义故宫南院和新北市立美术馆，仍然在自由与秩序之间保持了某种精妙的平衡。逸如流水，乐而不淫，放而不纵。姚仁喜拈逸为雅魄，得以令设计破匠入艺、超艺成道，使建筑阴阳化趣，气韵生动。

释家之"寂"，是深谙佛理的姚仁喜在设计中常常不自觉流露的一种气质，让他的建筑与热闹浓烈的尘世造物拉开了明显的距离。且不说养慧学苑、法鼓山农禅寺之类的释家道场，纵使如新竹高铁站、乌镇大剧院这样喧嚣扰攘的人间俗地，姚仁喜也依然展现出在物质表达上的高度克制——微微着意、淡淡呈示，既不热切，也非冷峻，而仅有对幻象终必成虚的悲悯和不得不在无住

之中建造的平静。姚仁喜营造的雅境在律宗和禅宗之间，或谓"律中见禅"：刹那芳华，于短暂中搭造永恒，于秩序间放下执着；正如克缇办公大楼在信义计画区的孤独显相——似近实远，似动还静，似真亦幻，一道入时的纱幔遮裹住不动声色的本来面目，任潮起潮落历遍周遭，蝉翼之隔却终难揭破。

清人邓石如有联："春风大雅能容物，秋水文章不染尘"；《礼记·中庸》有云："万物并育而不相害，道并行而不相悖。"春风大雅，是对万物并育、诸道并行的温煦照拂、侧耳倾听与包容和合。姚仁喜的设计，汇中西现代建筑之道于一体，熔儒道释三家之妙于一冶，删繁就简，弃巧归真，终化春风大雅为秋水文章。

La Biennale di Venezia 8th International Architecture Exhibition
NEXT EXIT - Hsinchu High Speed Rail Station
第八届威尼斯建筑双年展 - 下一出口：以"新竹高速铁路站"展出

La Biennale di Venezia 14th International Architecture Exhibition – Fundamental
Collateral Event - Time Space Existence - The Ninth Column
第十四届威尼斯建筑双年展 - "本源"大会平行展 - 时间·空间·存在 - 第九柱

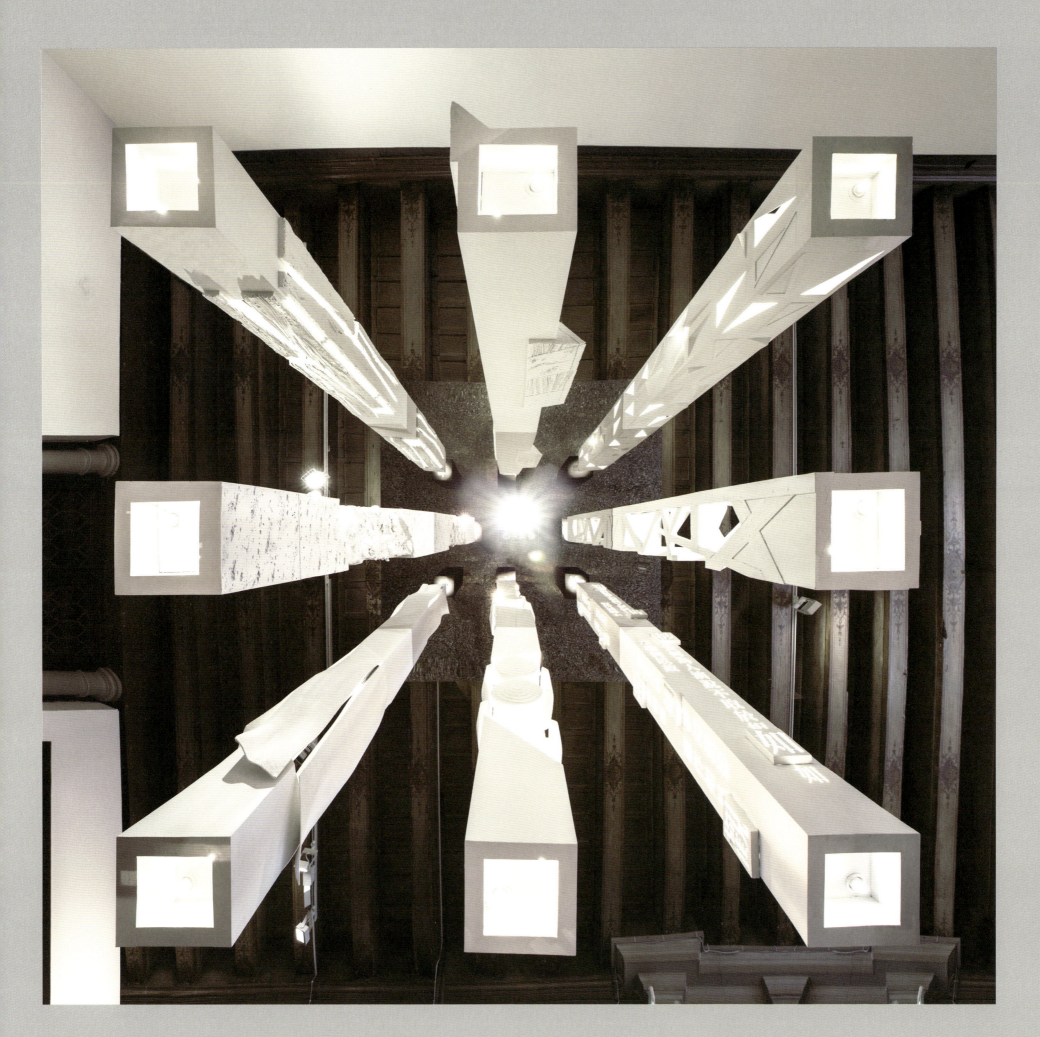

Community 聚 Forums

ARCHITECTURE AS THEATRICAL STAGE

Kris Yao

Architecture should neither be used only for mere practical functions, nor should it be used exclusively as a financial instrument or marketing tool. Architecture should strive to go beyond the singular function of an edifice, and become a stage for human living.

Buildings exist for the communal aggregation of people; hence, they are a theatrical stage of public activities. We as humans, in the buildings, are at times the audience and at others, an actor, performing for and watching one another. Architecture should celebrate the communal presence of people gathering in it, enabling us to freely and consciously play our respective roles in our lives from one theater to the next.

I like to create and shape the "scenes" in which people's lives unfold — a train station for example, where travelers scurry about their journeys while sharing an ephemeral connection that binds them together — a moment of transient, anxious solitude that permeates the space. Or in a bustling city plaza, a solitude yet communal sense shared by the collection of loosely gathered crowds. The architecture that supports and contains these human emotions should not exist independently just as an expression of professional or technical execution; architecture should be shaped through thorough care and empathy of humanity.

建筑是舞台

姚仁喜

我认为建筑不仅止于实用的功能而已，更不应沦为市场策略或财务投机的工具。建筑应该要超越单一的功能性机制，成为人们生活的舞台。

许多建筑是为了人的聚合而存在的，因此它应该是众人活动的剧场。我们在其中，既是观众，又是演员，相互参与、相互观赏、相互演出。建筑应该彰显场域精神并支持社会脉动，让人们自觉地扮演各自的角色，自在地生活在此剧场中。

我喜欢经营与创造人们生活的"场景"。例如，在车站中，形色匆匆的旅人们有着共同的情感——瞬变、不安、孤寂。或在都市的广场中，三三两两的市民对于享有既独立又群聚的期待。提供包容这些情感的建筑不只是某种专业或技术的独立存在，而是要对人们的共通性有相当的观照与同理心。

CONTINENTAL ENGINEERING CORPORATION HEADQUARTERS

汉德大楼

Taipei, Taiwan, China | Completion 1999
中国 台湾 台北 | 1999 年完工

Second floor lobby
二楼大厅

LOCATION	Taipei, Taiwan, China
CLIENT	Han-de Construction Co., Ltd.
FLOOR LEVELS	13 Floors, 4 Basements
BUILDING STRUCTURE	Steel Structure, Steel Reinforced Concrete Columns
MATERIALS	Architectural Concrete, Aluminum Panel, Granite, Gray Tinted-glass Unit
BUILDING USE	Office
SITE AREA	2114 ㎡
TOTAL FLOOR AREA	17572 ㎡
DESIGN INITIATIVE	1994
COMPLETION	1999

项目位址	中国 台湾 台北
业主	汉德建设股份有限公司
楼层	地上13层、地下4层
建筑结构	钢骨结构、钢骨钢筋混凝土柱
材料	清水混凝土、铝板、花岗岩、灰色复层玻璃
用途	办公大楼
基地面积	2114 ㎡
总楼地板面积	17572 ㎡
设计起始时间	1994
完工时间	1999

The Continental Engineering Corporation's company image and culture are reflected through its headquarter building design, crafted with sophisticated attention to detail by using the most basic building materials, namely concrete, steel and glass. The building is composed of a transparent glass box, eight reinforced concrete columns, exposed steel bracings at its four corners and a service core in the rear. The exoskeleton element strongly expresses the dynamics of the building structure: the mega-columns and floor trusses support the gravity loadings, and the exterior bracings resolve the lateral system for seismic loading. The technical challenges for design and construction of the exposed structural system, long span steel members and pour-in-place architectural concrete were not only been overcome collaboratively by the client, contractor and the design team, but also done so with elegance and grace.

汉德大楼的设计充分反映其作为营建公司的形象与文化。以最基本的建筑材料，即混凝土、钢材及玻璃，精密细致地雕琢出此建筑量体。建筑物由一方形玻璃盒、八组钢筋混凝土柱、四个角隅的外露钢构斜撑以及背面的服务核所组成。外露的结构骨架强烈表现出此建筑设计的力学：巨柱与地板桁架支撑了垂直重力，而外部斜撑则解决了水平结构系统的抗震。本建筑物的外露结构系统、大跨度钢构以及现场浇注清水混凝土，不论在设计或施工上皆是极大的技术挑战；但在业主、承包厂商与设计团队的紧密合作之下，不仅一一完成，亦使本案最终以优美而典雅的面貌呈现。

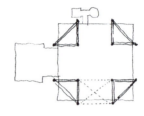

North facade at night
北側立面夜景

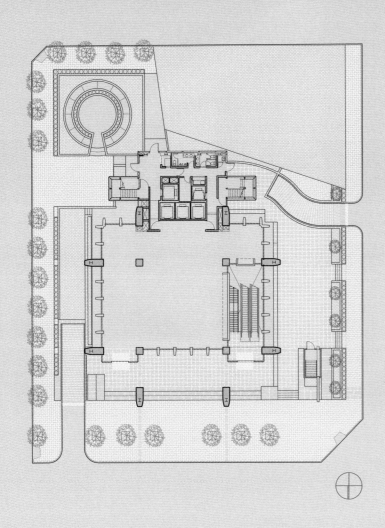

Ground floor plan

地面层平面图

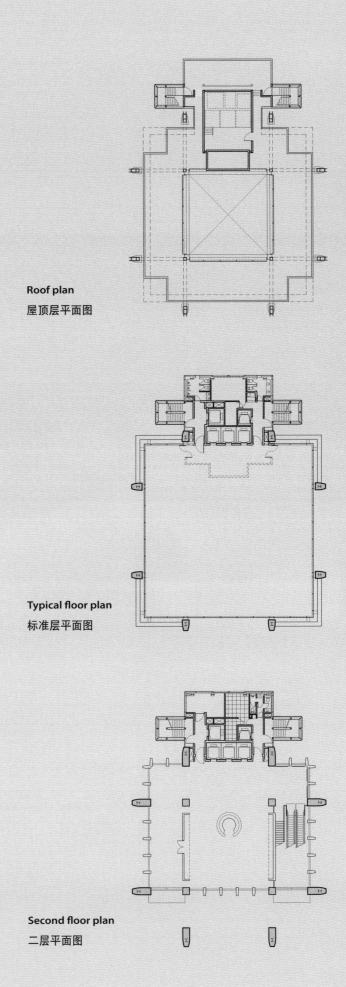

Roof plan

屋顶层平面图

Typical floor plan

标准层平面图

Second floor plan

二层平面图

Entrance plaza sculpture
入口广场雕塑

Second floor hallway
二楼走道

Facade detail
一楼局部立面

Escalator to second floor lobby
通往二楼大厅之电扶梯

Stair and glass wall details
楼梯及玻璃墙面细部

Details of the exposed structure
外露结构细部

QUANTA RESEARCH & DEVELOPMENT CENTER

广达研发中心

Taoyuan, Taiwan, China | Completion 2004
中国 台湾 桃园 | 2004 年完工

LOCATION	Taoyuan, Taiwan, China
CLIENT	Quanta Computer Inc.
FLOOR LEVELS	7 Floors, 3 Basements
BUILDING STRUCTURE	Steel Structure
MATERIALS	Perforated Aluminum Panel, Low-E Glass, Fritted Tempered Glass, Granite, Teflon Membrane
BUILDING USE	Research & Development Center
SITE AREA	48311 ㎡
TOTAL FLOOR AREA	212371 ㎡
DESIGN INITIATIVE	2002
COMPLETION	2004

项目位址	中国 台湾 桃园
业主	广达科技股份有限公司
楼层	地上 7 层、地下 3 层
建筑结构	钢骨结构
材料	冲孔铝板、低辐射玻璃、彩釉网点强化玻璃、花岗岩、铁氟龙薄膜
用途	研发中心
基地面积	48311 ㎡
总楼地板面积	212371 ㎡
设计起始时间	2002
完工时间	2004

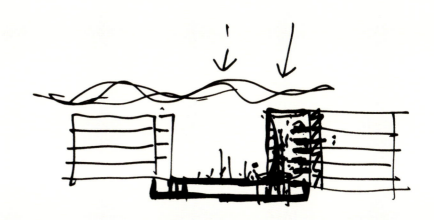

The Quanta Research and Development Center is a high-tech campus for technological research, cultural and leisure functions. Apart from R & D spaces, the complex also houses a performing hall, a museum, a "Think Tank" and staff amenities such as recreation and dining facilities. The complex is composed of two interlocking masses resonate with the traditional Chinese *Ying* and *Yang*, leaving open a large courtyard that serves as the focal point for the complex.

The canopy covering the courtyard is the most prominent element of the center above the inner plaza. The aerodynamic "Think-Tank" projects outward from the canopy offering an expansive view of the center and the grand canopy below. The east, west and south facades of the building are adorned with barcode-like aluminum and glass sunshades. The perforated aluminum panels and patterned glass create a visually appealing layered elevation with a high-tech image of the building.

广达研发中心这座复合式建筑物包含了科技研发、文化设施以及员工休闲功能。除了主要的研发空间之外，还包括了演艺厅、博物馆、智库空间以及员工的休闲与餐饮设备等设施。整体建筑物由两个互相环扣的量体所构成，象征中国传统的"阴阳合德"。中央宽广的中庭，成为这栋复合建筑物的焦点。

覆盖中庭的顶篷，是建筑物设计最显著的中心要素，弧面造型与透明材质，宛如广场上空飘浮的白云。造型顺应空气动力学的"智库空间"从顶篷向上延伸，像遨翔在云上的龙头俯视下方，可以从此处透过10米高的透明玻璃帷幕，饱览下方园区及大顶篷的丰富景观。东、西、南面的建筑立面以条码图像的玻璃与铝制遮阳板构成。冲孔铝板与陶版印花玻璃创造出富层次感的立面设计，营造高科技的建筑意象。

Longitudinal section
纵向剖面图

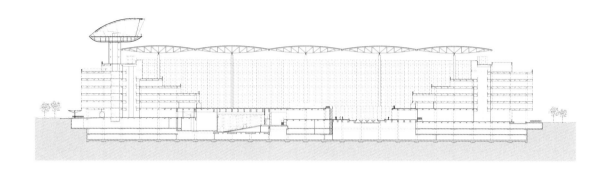

Second floor plan
二层平面图

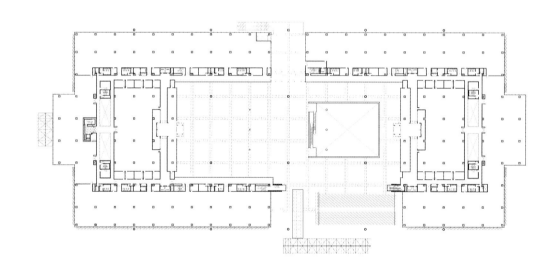

Ground floor plan
地面层平面图

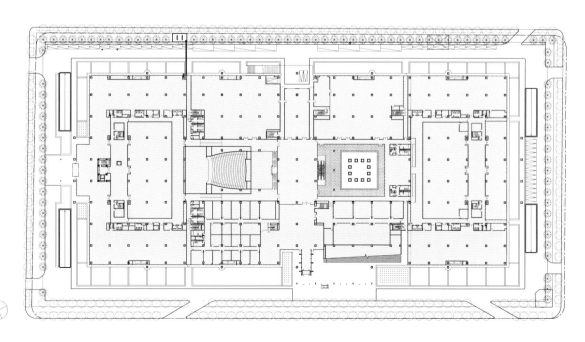

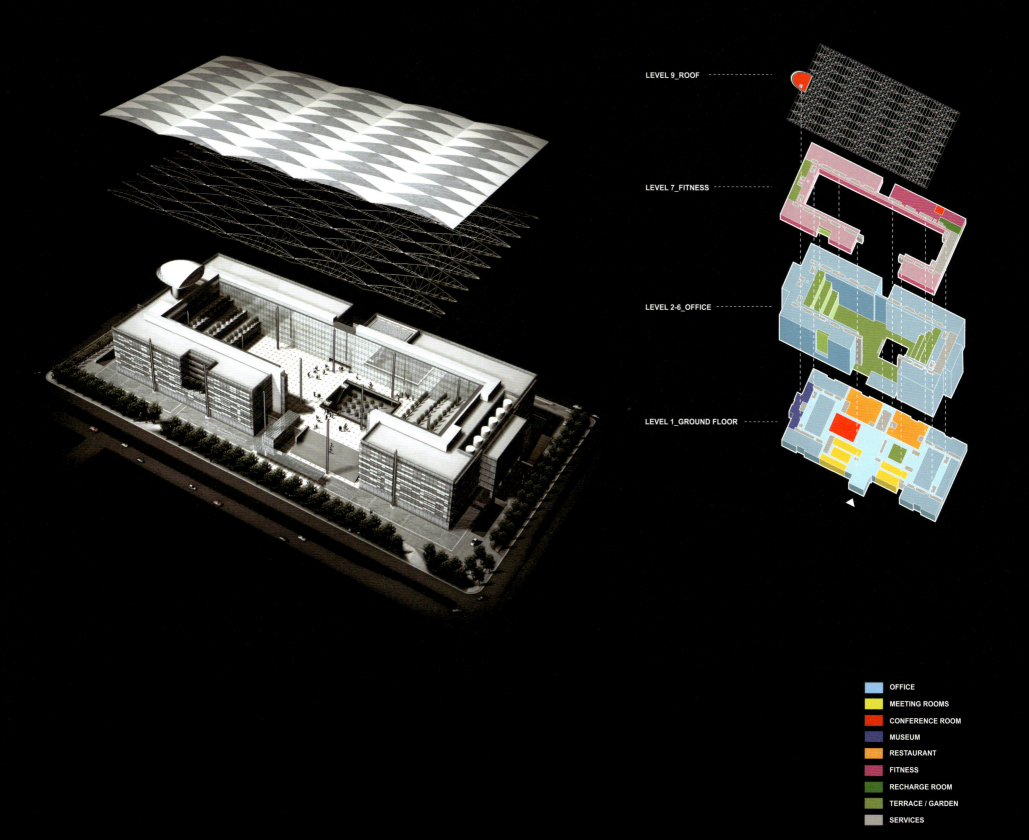

Grand canopy and courtyard
大顶篷及中庭广场

广达研发中心

View from the Think-Tank
智库空间窗景

Concept model of the Think-Tank
智库空间概念模型

Quanta Concert Hall
广艺厅

Acoustic wall detail
音效墙面细部

KELTI CENTER

克缇大楼

Taipei, Taiwan, China | Completion 2009
中国 台湾 台北 | 2009 年完工

LOCATION	Taipei, Taiwan, China
CLIENT	Jing Yung Gi Investment Co., Ltd.
FLOOR LEVELS	15 Floors, 3 Basements
BUILDING STRUCTURE	Steel Structure
MATERIALS	Low-E Glass, Champagne-color Aluminum Panel, Granite
BUILDING USE	Office
SITE AREA	5864 ㎡
TOTAL FLOOR AREA	28823 ㎡
DESIGN INITIATIVE	2005
COMPLETION	2009

项目位址	中国 台湾 台北
业主	金永基股份有限公司
楼层	地上 15 层、地下 3 层
建筑结构	钢骨结构
材料	低辐射玻璃、香槟色烤漆铝板、花岗岩
用途	办公大楼
基地面积	5864 ㎡
总楼地板面积	28823 ㎡
设计起始时间	2005
完工时间	2009

The Kelti Center sits at the end of a major street in the busy downtown Taipei. Its majestic form, delicate crystal-like glass volumes and the generous and beautifully crafted plaza, enabled the building to anchor the site and become a prominent landmark in this area.

Two low-rise volumes flanked by the service cores on each side and a high-rise crystal glass volume suspended from the "portal frame" make up the entire complex. The space between the lower volumes forms the building's entrance. The building's lower commercial levels are designated for showrooms and restaurants. The middle section houses the corporate offices, and the top floor is a sky-lit, 13.5-meter-high event hall with a spectacular view to the city. Natural sunlight streams into the restaurants in the basement from the expansive first floor lobby. The service cores are located on the north and south sides to give maximum flexibility for the spaces in between.

克缇办公大楼位于台北市区繁荣商圈的要道底端。雄伟的造型、黑水晶般的玻璃量体，加上建筑物前方留设宽阔的广场，使得克缇大楼坚稳地锚定于此街道的端点，成为该区域重要的地标建筑。

整体建筑由门型框架、夹在其间的两座低层量体以及悬挂在上方的水晶般玻璃量体所构成。两座低层量体之间形成大楼的主入口。如结晶岩石般的玻璃量体以复层深色低辐射玻璃构成，门型框架则以水冲式及平光面处理的石材包复，开创出新颖韵律对比，赋予高雅美学风格。建筑物低楼层为展示中心与餐厅，中间区段为企业办公室，顶楼是13.5米高的无柱活动会馆，引入天光，可饱览市区视野景观。位于地下室的餐厅享有自一楼入口大厅流泻而下的自然光线。南北两侧的服务核为整体建筑物带来最大的空间灵活弹性。

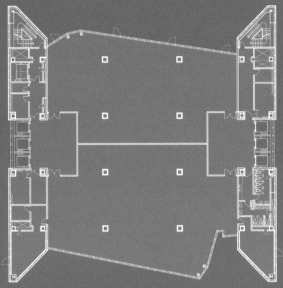

Fourth floor plan
四层平面图

Eleventh floor plan
十一层平面图

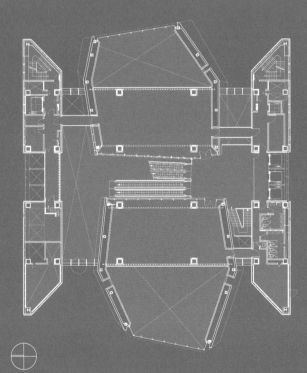

Second floor plan
二层平面图

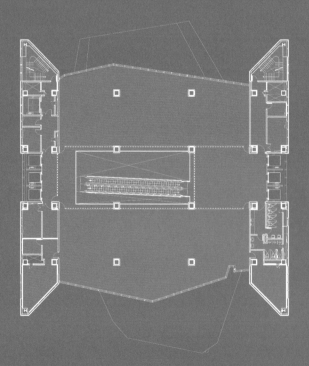

Third floor plan
三层平面图

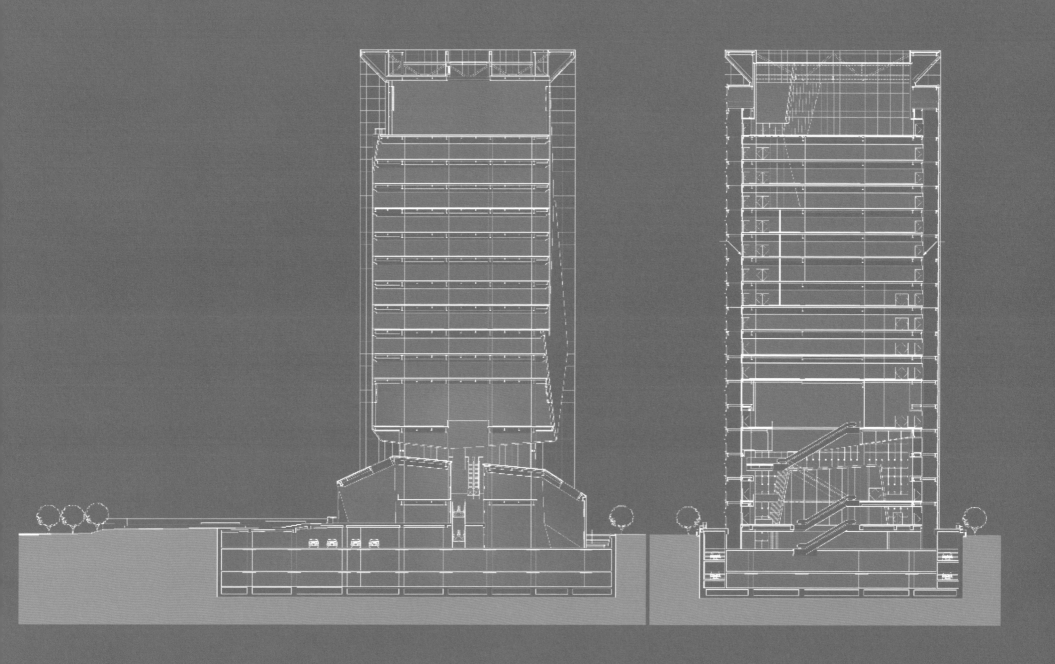

East-west section
东西向剖面图

North-south section
南北向剖面图

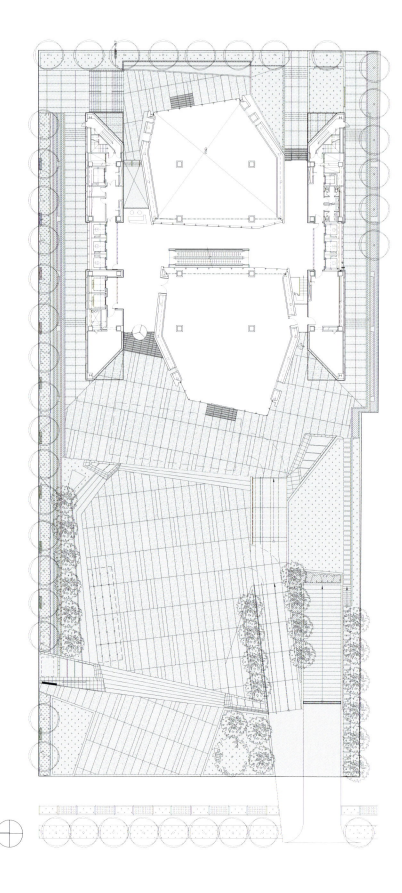

Site plan
基地配置图

View to the landscape from second floor atrium
二楼挑空区外窗景

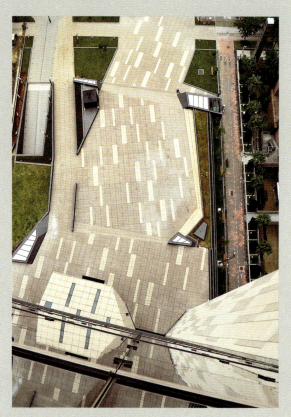

View to the entrance plaza
一楼入口广场俯视

East facade and park
东向立面及公园

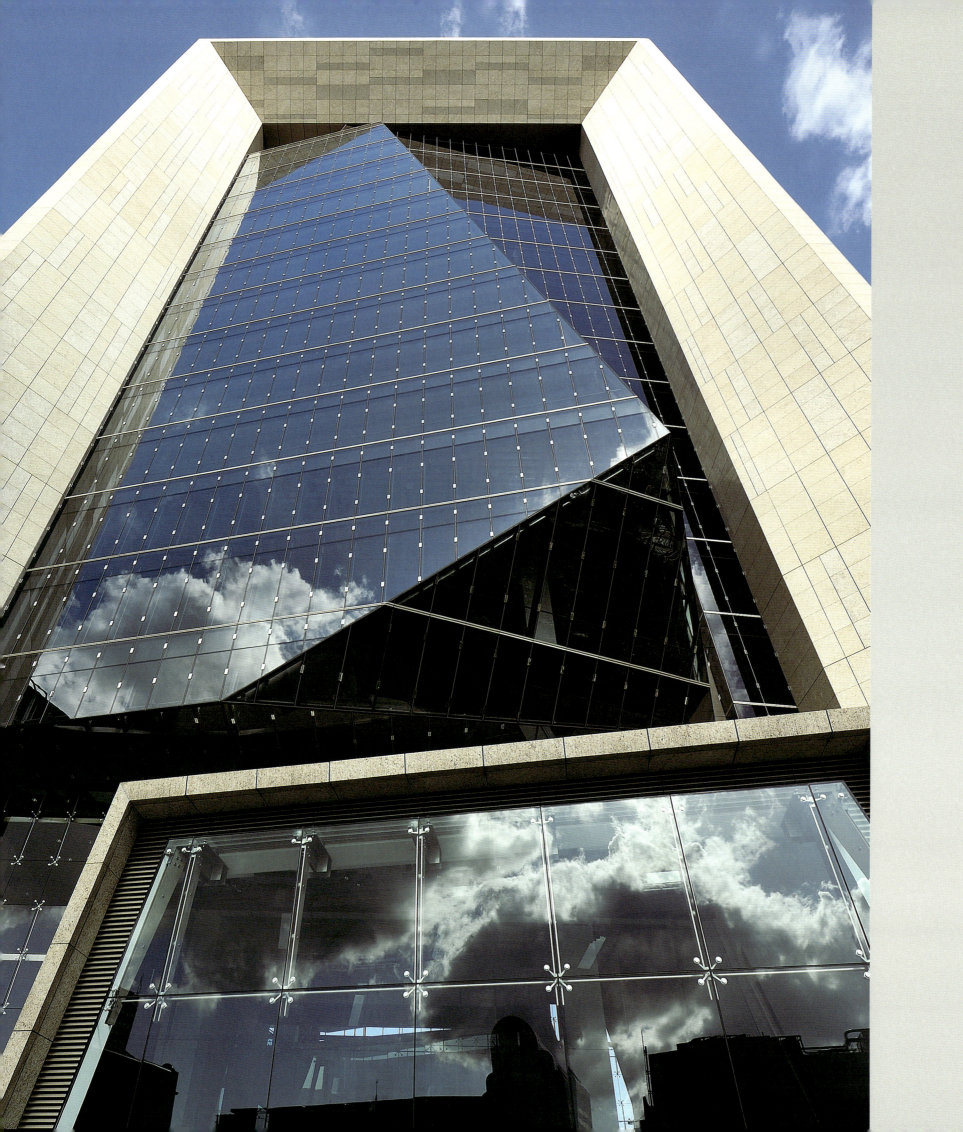

克缇大楼 63

Roof terrace at top floor
顶层露台

Eleventh floor private terrace
十一层平台

Ground floor lobby
一楼大厅

克缇大楼　71

HUA NAN BANK HEADQUARTERS

华南银行总部

Taipei, Taiwan, China | Completion 2014
中国 台湾 台北 | 2014 年完工

LOCATION	Taipei, Taiwan, China
CLIENT	Hua Nan Commercial Bank Ltd.
FLOOR LEVELS	27 Floors, 2 Basements
BUILDING STRUCTURE	Steel Structure
MATERIALS	Low-E Glass, Extruded Aluminum Louver, Aluminum Panel, Stainless Steel, Granite
BUILDING USE	Office
SITE AREA	8945 ㎡
TOTAL FLOOR AREA	51881 ㎡
DESIGN INITIATIVE	2008
COMPLETION	2014

项目位址	中国 台湾 台北
业主	华南商业银行股份有限公司
楼层	地上 27 层、地下 2 层
建筑结构	钢骨结构
材料	低辐射玻璃、铝挤型百叶、铝板、不锈钢、花岗岩
用途	办公大楼
基地面积	8945 ㎡
总楼地板面积	51881 ㎡
设计起始时间	2008
完工时间	2014

The 154.5 meters tall Hun Nan Headquarters is comprised of three-story podium and twenty–seven-story office tower, with the main entrance and long side of the rectilinear plan facing the west. Detached from the exterior curtain wall, the peripheral structure is pushed out as an exoskeleton system, allowing the spaces in between for balconies, which also act as sun-shading devices to enhance sustainability in design. This design feature allows the interior space to be column-free, with a clean and smooth window line around it for flexibility in space planning. On the front center bay of the tower, a series of double-height "sky gardens" are stacked all the way to the top of the tower. They make for casual meeting and relaxing spaces for the office staff, and also serve as the buffer zones for reducing heat-gain from the west facade. Oversize ceiling fans are installed in these spaces to achieve the thermal destratification effect.

With additional energy saving strategies applied throughout the building system, this bank headquarters has acquired a LEED Gold rating in sustainable design.

华南银行总部由三层裙楼与二十七层办公塔楼所组成，总高 154.5 米。建筑物正面朝西。设计将外围结构向外推出，形成外露结构系统。此结构与帷幕墙分离，其间局部作为阳台，也同时作为遮阳，强化环保的永续性。此设计不仅让建筑物内完全无柱，更让外缘窗线平整流畅，提供更多灵活运用的空间弹性。建筑量体前方的中央凹处设置有一系列双层高的空中花园，除了可供员工平常会谈、休憩之用，还能作为减缓西晒的缓冲区。这些空间的天花板上设置了超大型的风扇，以扰流效应达到空调节能的效果。

华南银行总部获得 LEED 绿建筑金质标章的认证。

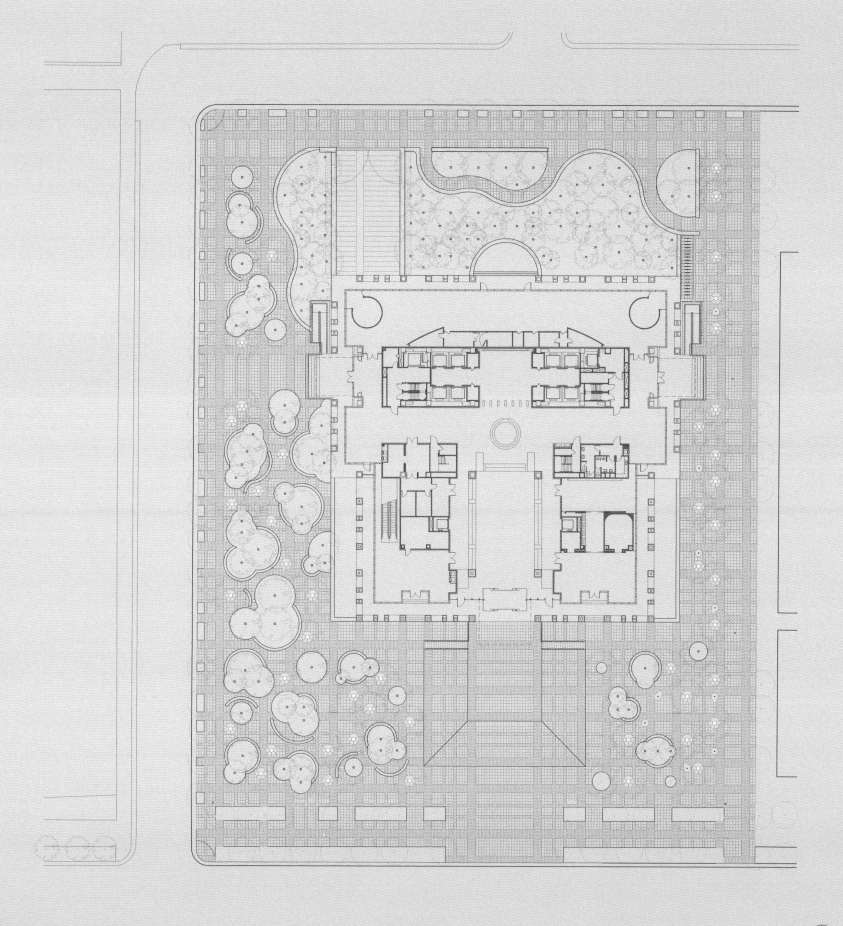

Ground floor plan
地面层平面图

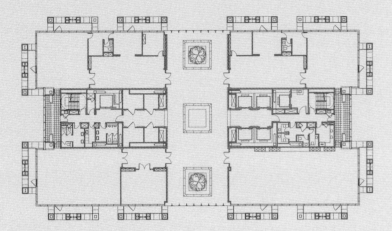

Twenty-seventh floor plan

二十七层平面图

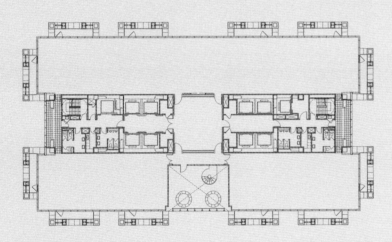

Typical floor plan

标准层平面图

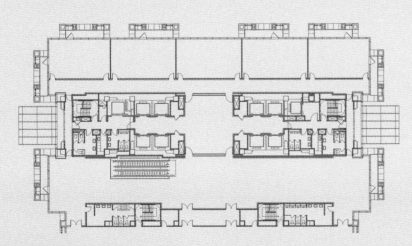

Second floor plan

二层平面图

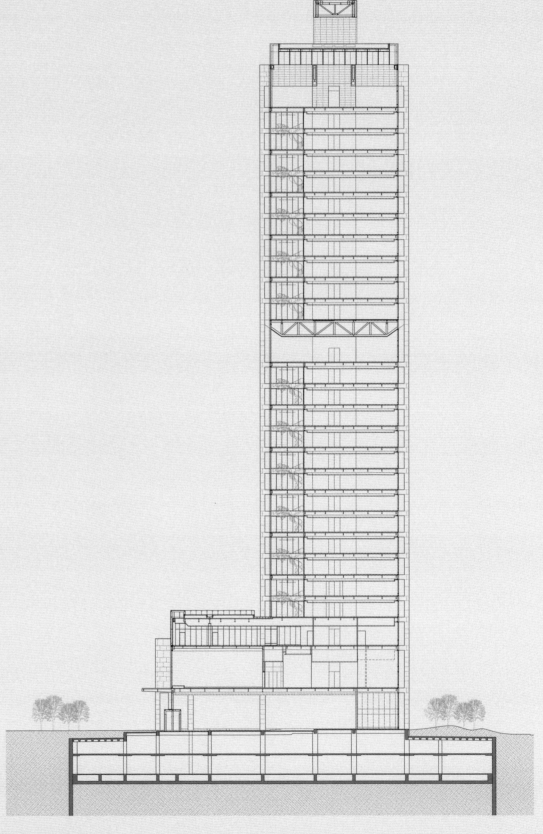

East-west section

东西向剖面图

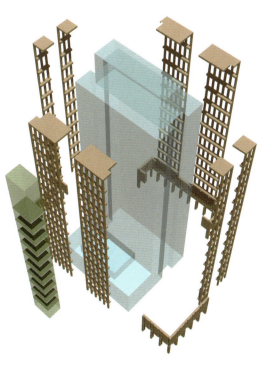

Exploded diagram
分解示意图

West facade
西向立面

The double-height sky garden
双层空中花园

The double-height sky garden
双层空中花园

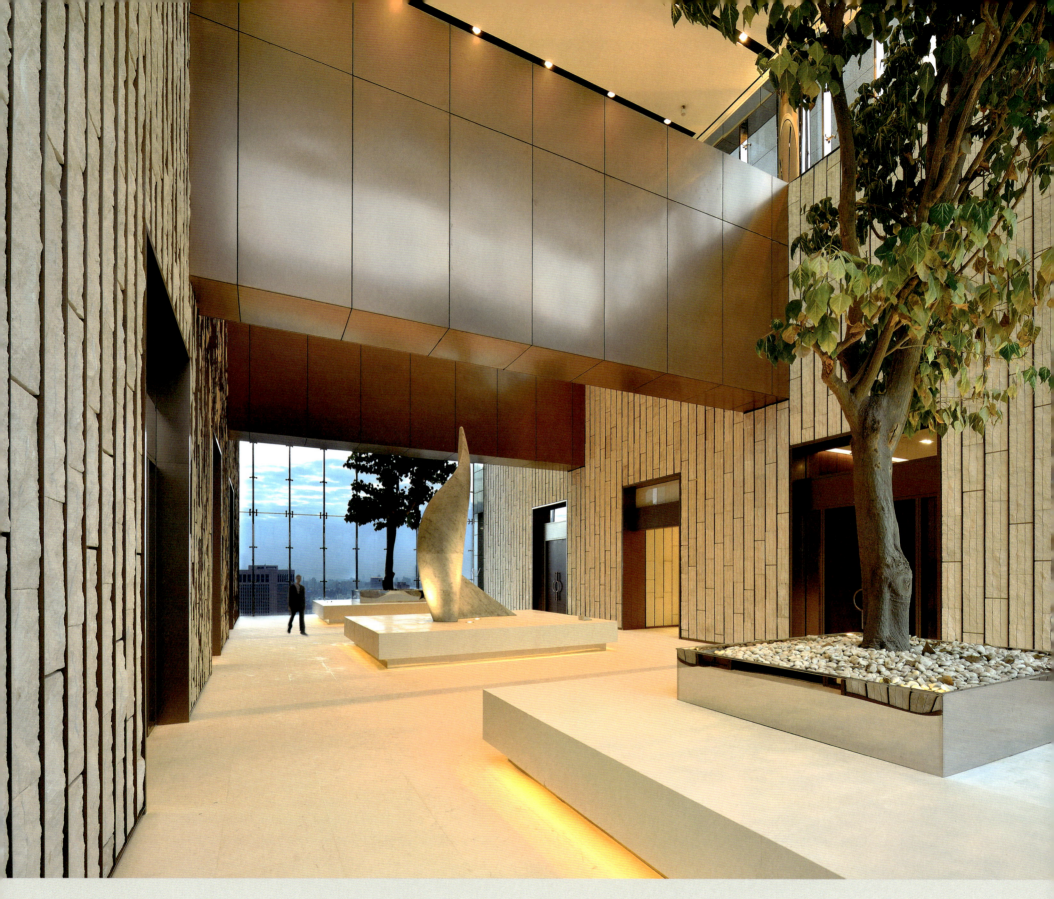

Executive Floor Lobby
主管层大厅

Entrance Lobby
入口大厅

中钢集团总部

Kaohsiung, Taiwan, China | Completion 2012
中国 台湾 高雄 | 2012 年完工

CHINA STEEL CORPORATION HEADQUARTERS

LOCATION	Kaohsiung, Taiwan, China
CLIENT	China Prosperity Development Corp.
FLOOR LEVELS	29 Floors, 4 Basements
BUILDING STRUCTURE	Steel Structure
MATERIALS	Double Skin Glazed Curtain Wall, Low-E Insulation Glass, Stainless Steel, Aluminum Panel, Granite
BUILDING USE	Office
SITE AREA	11037 ㎡
TOTAL FLOOR AREA	81054 ㎡
DESIGN INITIATIVE	2004
COMPLETION	2012

项目位址	中国 台湾 高雄
业主	中欣开发股份有限公司
楼层	地上29层、地下4层
建筑结构	钢骨结构
材料	双层玻璃帷幕墙、低辐射复层玻璃、不锈钢、铝板、花岗岩
用途	办公大楼
基地面积	11037 ㎡
总楼地板面积	81054 ㎡
设计起始时间	2004
完工时间	2012

The port city Kaohsiung located in southern Taiwan is transforming itself from an industrial town to a multi-functional business-trading city. The China Steel Corporation Headquarters is situated in an area adjacent to the port, where the largest urban development in Kaohsiung in recent years is taking place. This area will facilitate functions ranging from transportation, logistics and trading to culture, recreation and institution. The headquarters will be an integral element of this new area, and will also become the new landmark for the Kaohsiung Port.

The building is composed of four square tubes bundled together by a shared central core. Each tube is turned 12.5 degrees in increments of eight stories, forming a dynamic geometry. The exterior mega-bracings span every eight stories, with their tiebacks forming terraces on every interval. The diamond-shaped double skin curtain wall allows for optimized natural lighting and ventilation that reduces heat-gain, minimizes energy consumption and shields traffic noise in this warm urban climate. On the ground level, the square tower sits on a round water pond at the center of the site. The remaining site is densely landscaped with trees to provide a comfortable pedestrian environment.

台湾南部的高雄港正在从工业城转型为多功能经贸园区，而中钢企业总部大楼所在的临港地区，正是近年高雄市最大型的都市建设催生地，功能内容从运输、物流、经贸，一直到文化、休闲与公共机构等方面。这栋总部大楼为此新兴区中不可或缺的重要元素，也已成为高雄港的新门户地标。

整栋建筑以四个方形量体包夹着中央轴心服务核所构成。方形量体以八层楼为一单元，扭转 12.5 度，因而形成生动的几何形体。外观上，大型结构斜撑每次跨越八层楼，并且在相连结处形成大阳台。钻石形状的双层帷幕墙提供了最佳自然光线及良好通风，在亚热带的都会气候里，具有阻热、节能、降低噪音等优点。方形塔楼在地面层优雅地矗立于圆形水池中央；其余部分则种植许多花草树木予以美化，为行人建构出绿意盎然的舒适环境。

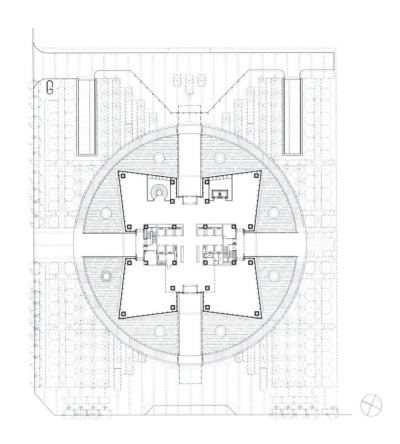

Ground floor plan
地面层平面图

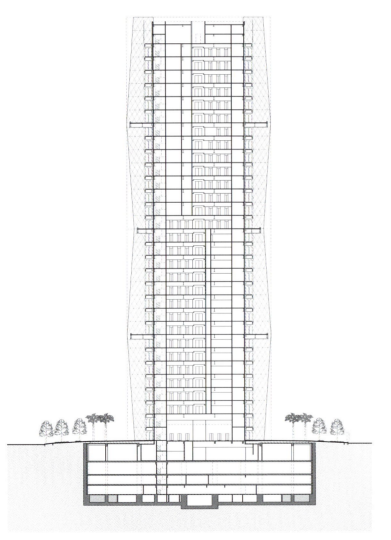

East-west section
东西向剖面图

中钢集团总部

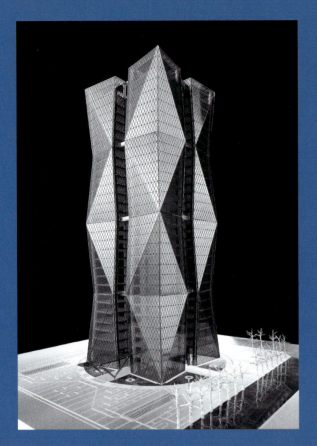

Model, view of northeast
模型，东北侧角度

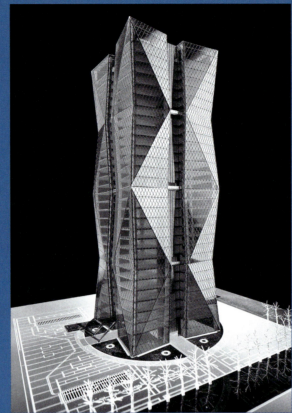

Model, entrance view
模型，入口立面

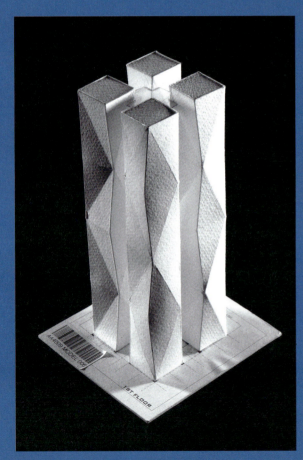

Concept model
概念模型

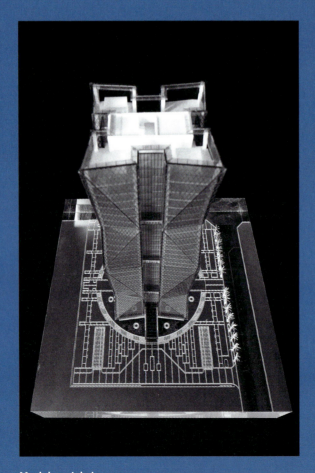

Model, aerial view
模型，鸟瞰

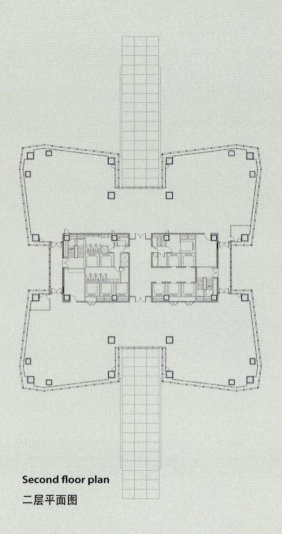

Second floor plan
二层平面图

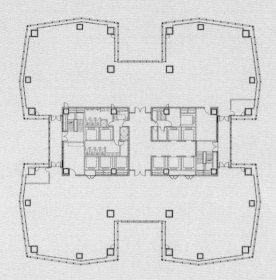

Fifth floor plan
五层平面图

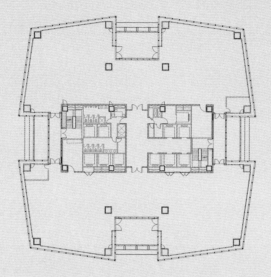

Eighth floor plan
八层平面图

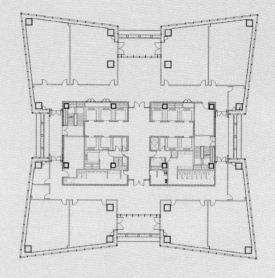

Sixteenth floor plan
十六层平面图

Entrance canopy and landscape pool
入口雨遮及景观水池

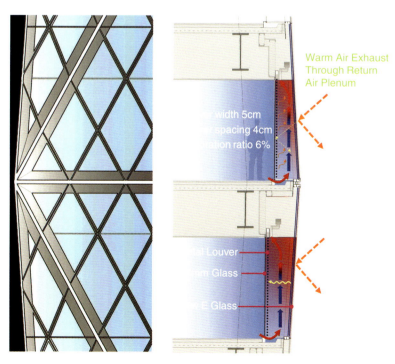

Air flow diagram of the double-skin curtain wall
双层帷幕墙系统，空气流动示意图

Corner of ground floor lobby
室内大厅一角

Reception counter
接待柜台

Facade detail at night
局部立面夜景

Facade detail
局部立面

HUMBLE HOUSE

寒舍艾丽酒店

Taipei, Taiwan, China | Completion 2013
中国 台湾 台北 | 2013 年完工

North facade at night
北向立面夜景

LOCATION	Taipei, Taiwan, China
CLIENT	Fubon Life Insurance Co., Ltd.
FLOOR LEVELS	22 Floors, 5 Basements
BUILDING STRUCTURE	Steel Structure
MATERIALS	Aluminum Panel, Low-E Glass, Insulated Glass, Corrugated Glass, Perforated Aluminum Panel
BUILDING USE	Hotel, Shopping Center
SITE AREA	6373 ㎡
TOTAL FLOOR AREA	54956 ㎡
DESIGN INITIATIVE	2009
COMPLETION	2013

项目位址	中国 台湾 台北
业主	富邦人寿保险股份有限公司
楼层	地上 22 层、地下 5 层
建筑结构	钢骨结构
材料	铝板、低辐射玻璃、复层玻璃、波浪玻璃、冲孔铝板
用途	旅馆、商场
基地面积	6373 ㎡
总楼地板面积	54956 ㎡
设计起始时间	2009
完工时间	2013

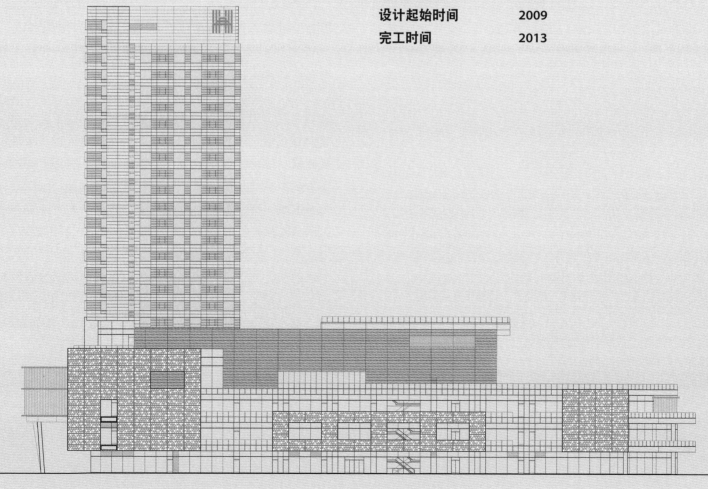

West elevation
西向立面图

Located in dense downtown Taipei, where high-rise commercial buildings are bridged by skywalks, the Humble House sits on one of the last remaining lots in the prominent Xinyi District, making it a key pedestrian hub within the district.

The project has three components: a 250-room business hotel in the main tower, a retail mall housed in the podium, and the parking and technical facilities in the third to fifth basement levels. The tower adapts a "wind-mill" plan to gain maximum views for its relative small guestroom units. Public venues such as restaurants, the ballroom, conference areas, spa and swimming pool are conveniently located right below the guest rooms, occupying the middle portion of the building with various levels of roof terraces. On the ground level, the retail mall and the hotel share a sizable plaza designed for passenger drop-off and occasional events.

On the exterior, sharp triangular bay window units gives an interesting rhythmic order to the tower and expand the view for the small rooms. For the podium, various cladding materials, including corrugated glass, perforated metal screens, transparent and translucent glass walls are applied, composing a vibrant facade that reflects the varied functions within.

寒舍艾丽酒店位于寸土寸金、人口稠密的台北市中心，高耸的商业大楼以空桥相连，借由寒舍艾丽酒店的完成，将信义区重要的街衢枢纽完全联系起来。

本案由三个部分组成：主楼是250间客房的商业酒店，裙楼设有购物商场，地下三至五楼作为停车场与机电设备。塔楼以风车状的平面配置，使得相对面积较小的客房也能享有最开阔的视野。客房正下方的中央楼层设有餐厅、宴会厅、会议室、水疗中心与游泳池等公共区域，配合着大小不一的屋顶露台。购物商场与酒店在地面层共用宽敞的入口广场，以方便旅客上下车或不定期举办活动之用。

酒店建筑外观的三角形广角窗，赋予整栋建筑饶富趣味的节奏感，同时也开展了小客房的视野。建筑裙楼采用不同外墙饰板，包括波浪玻璃、冲孔金属板、透明与半透明玻璃墙，构成视觉缤纷的外观，反映出内部多元的空间机能。

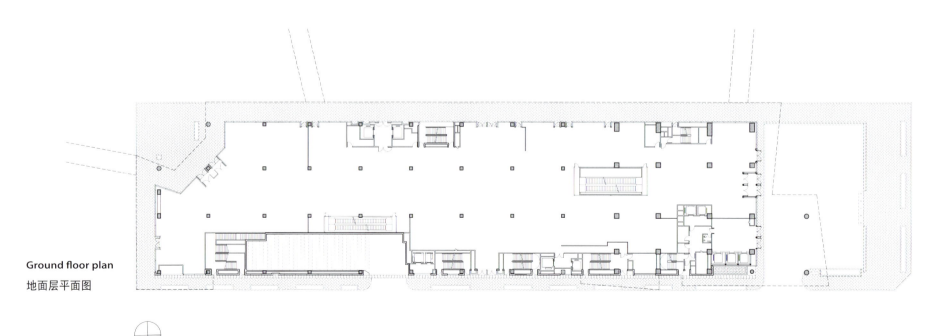

Ground floor plan
地面层平面图

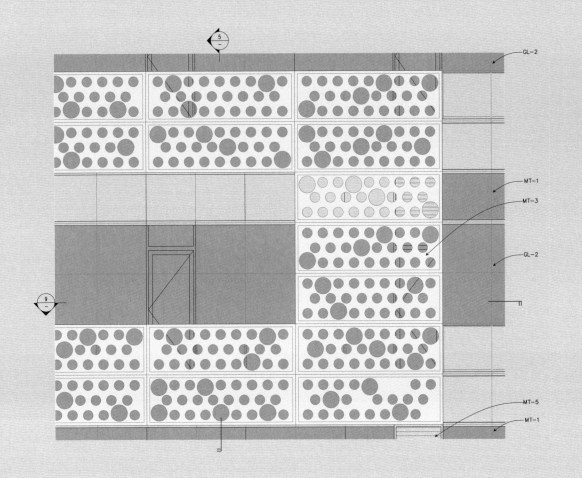
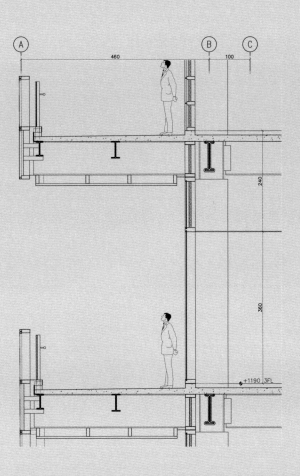

Section and elevation of the perforated aluminum screens
冲孔铝板剖立面图

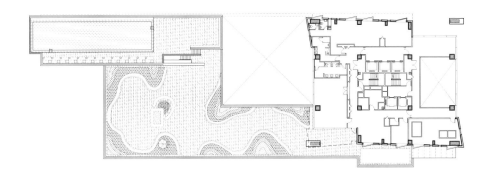

Sixth floor plan
六层平面图

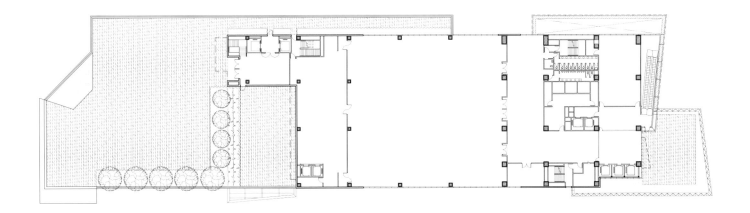

Fourth floor plan
四层平面图

寒舍艾丽酒店 115

Model, view of south
模型，南侧角度

Model, view of east
模型，东侧角度

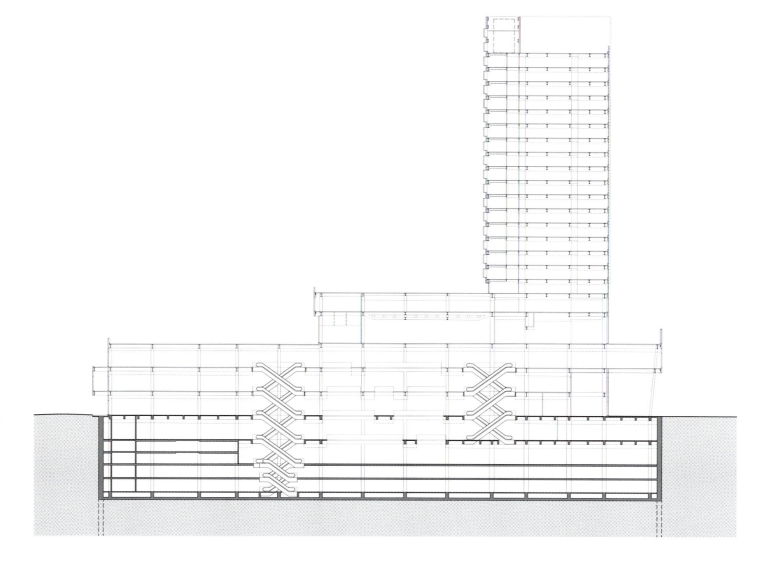

Longitudinal section
纵向剖面图

Model, view of podium
模型，裙楼

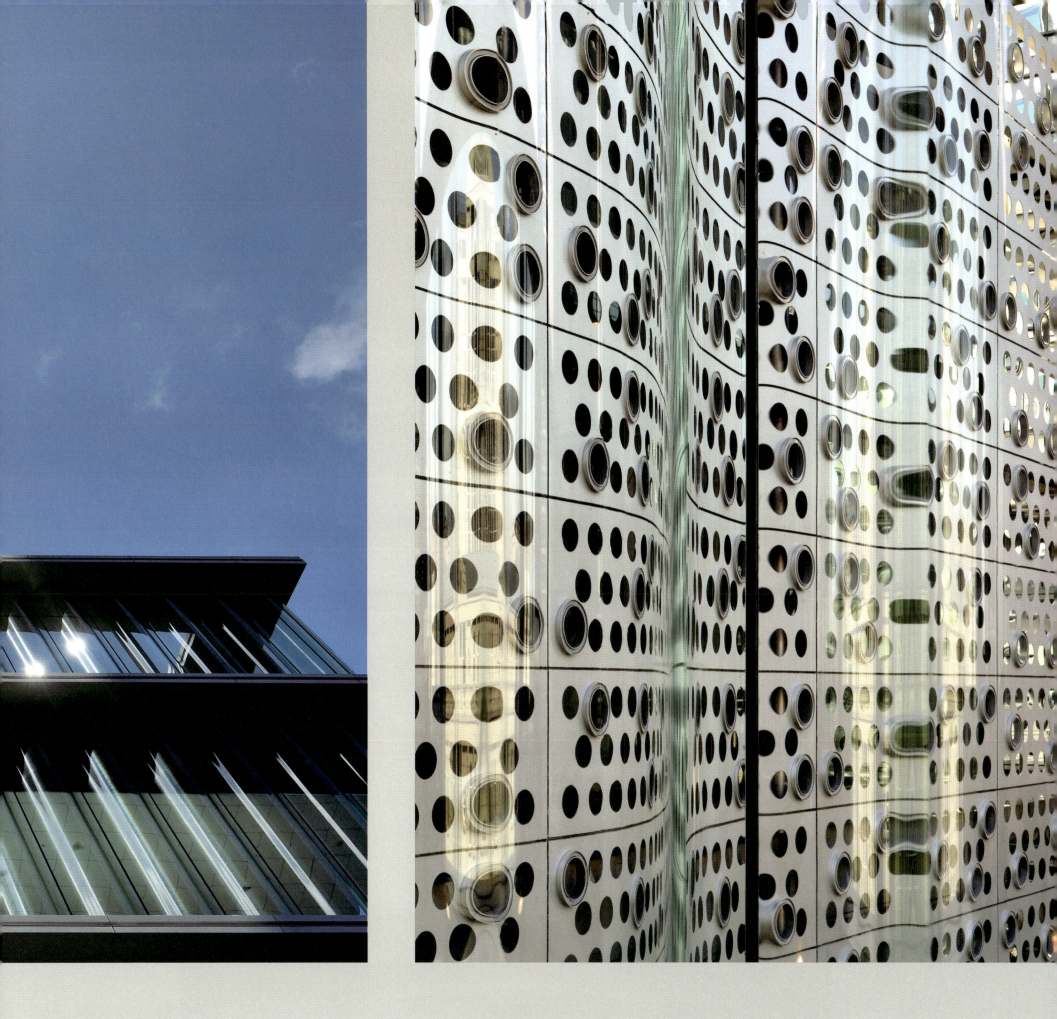

寒舍艾麗酒店

寒舍艾丽酒店 121

HSINCHU HIGH SPEED RAIL STATION

新竹高速铁路站

Hsinchu, Taiwan, China | Completion 2006
中国 台湾 新竹 | 2006 年完工

新竹高速铁路站

LOCATION	Hsinchu, Taiwan, China
CLIENT	Taiwan High Speed Rail Corporation
FLOOR LEVELS	3 Floors, 1 Basement
BUILDING STRUCTURE	Concrete Structure, Steel Roof Truss, Metal Roof Panel
MATERIALS	Architectural Concrete, Low-E Glass, Metal Panel, Stainless Steel
BUILDING USE	Station
SITE AREA	79900 m²
TOTAL FLOOR AREA	20361 m²
DESIGN INITIATIVE	2000
COMPLETION	2006

项目位址	中国 台湾 新竹
业主	THSRC 高速铁路股份有限公司
楼层	地上 3 层、地下 1 层
建筑结构	钢筋混凝土、钢骨屋顶桁架、金属板屋顶
材料	清水混凝土、低辐射玻璃、金属板、不锈钢
用途	车站
基地面积	79900 m²
总楼地板面积	20361 m²
设计起始时间	2000
完工时间	2006

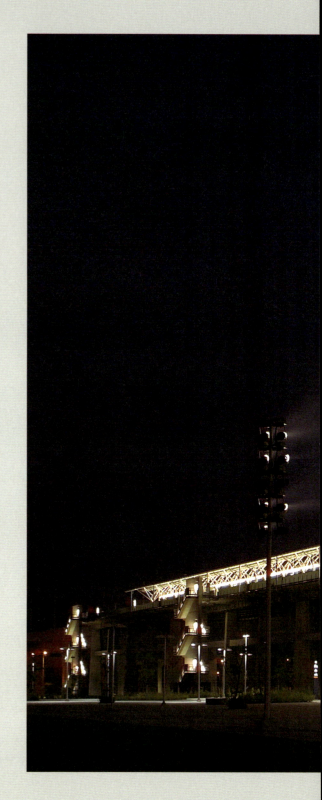

新竹高速铁路站

The Hsinchu High Speed Rail Station serves as a gateway for passengers arriving in the Hsinchu metropolis. The station's roof is designed in response to Hsinchu's infamous strong seasonal winds, and its curved roof looks like a billowing sail; from farther away, the slender, graceful roof appears like a diagonally folded sheet of paper. Sculpted with the motion of a dancer's tension-filled limbs in mind, it clearly shows the architect's earliest instant of inspiration.

The roof of the main structure is an arched surface formed by bending a parallelogram diagonally. It is supported via six sets of colossal trusses, and the two ends of the roof plane touch down lightly onto the ground, anchored to custom-designed pillars on either side of the station building, enhancing the soaring visual effect of its pneumatic form. The space for the tracks and platforms are of open-air design, eliminating the air compression produced when trains pass through at high speeds. The curved glass curtain walls and elongated stone walls form the symmetrical oval station concourse. The interior makes ample use of natural light sources, and two arc walls, one each at the northbound and southbound entrance, are adorned with glass sculptures respectively themed Future and Tradition, creating a dialogue between the new and the old.

新竹高速铁路站是旅客抵达新竹都会区的门户象征。新竹站的屋顶结构因应新竹地区著名的强大季节风而成型。曲面的屋顶仿佛被风鼓动的风帆，随着视线的拉远，轻盈薄透的量体如同一张对角弯折的纸片，状似舞者充满张力的肢体动线，清晰地表现出建筑师最初的灵感瞬间。

车站主体屋顶为平行四边形弯曲而成的曲面屋顶，屋面由六组巨型空间桁架撑起，两端以轻巧的姿态落地，锚定于车站主体两侧的造型柱上，强化视觉上飞跃动势的气体力学造型。车行轨道与月台空间为露天式设计，解决列车高速通过月台时产生的活塞效应。曲面玻璃帷幕及延伸的石材面墙形塑出对称与椭圆形的车站大厅。室内大量采用自然光源，两道弧形墙面分别位于北上与南下的入口，以艺术性的创作呈现，成为"未来"与"传统"两者之间新旧的对话。

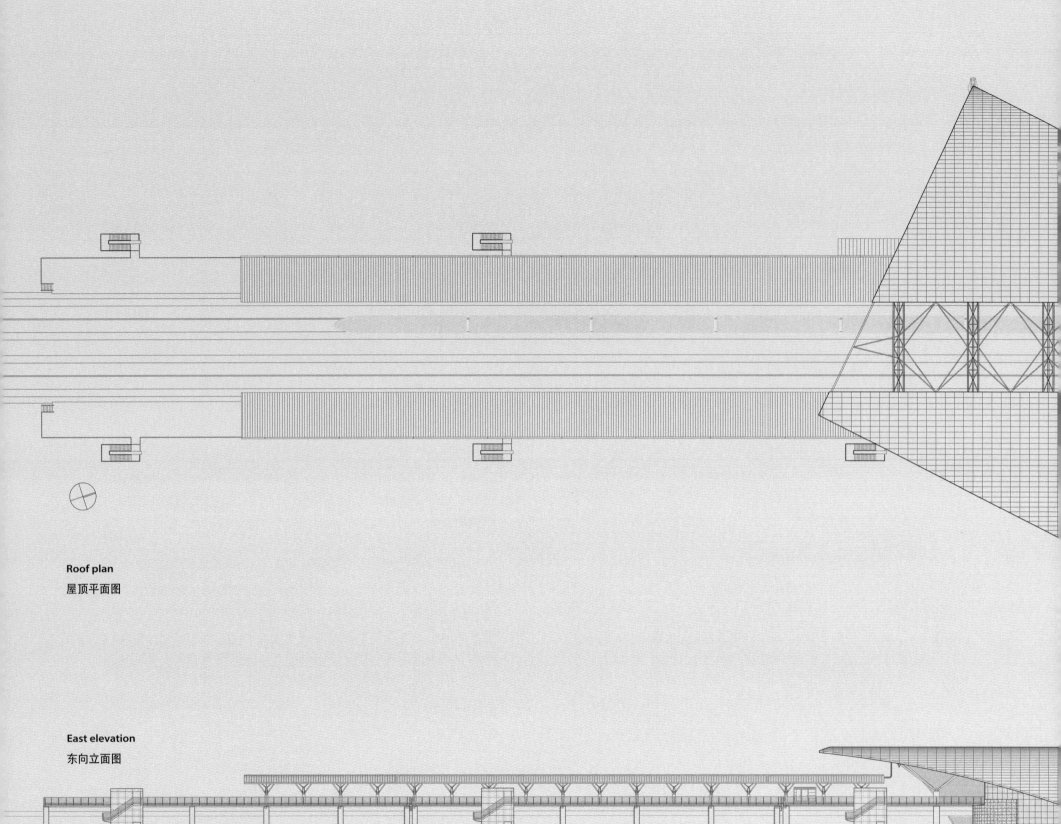

Roof plan
屋顶平面图

East elevation
东向立面图

新竹高速铁路站 131

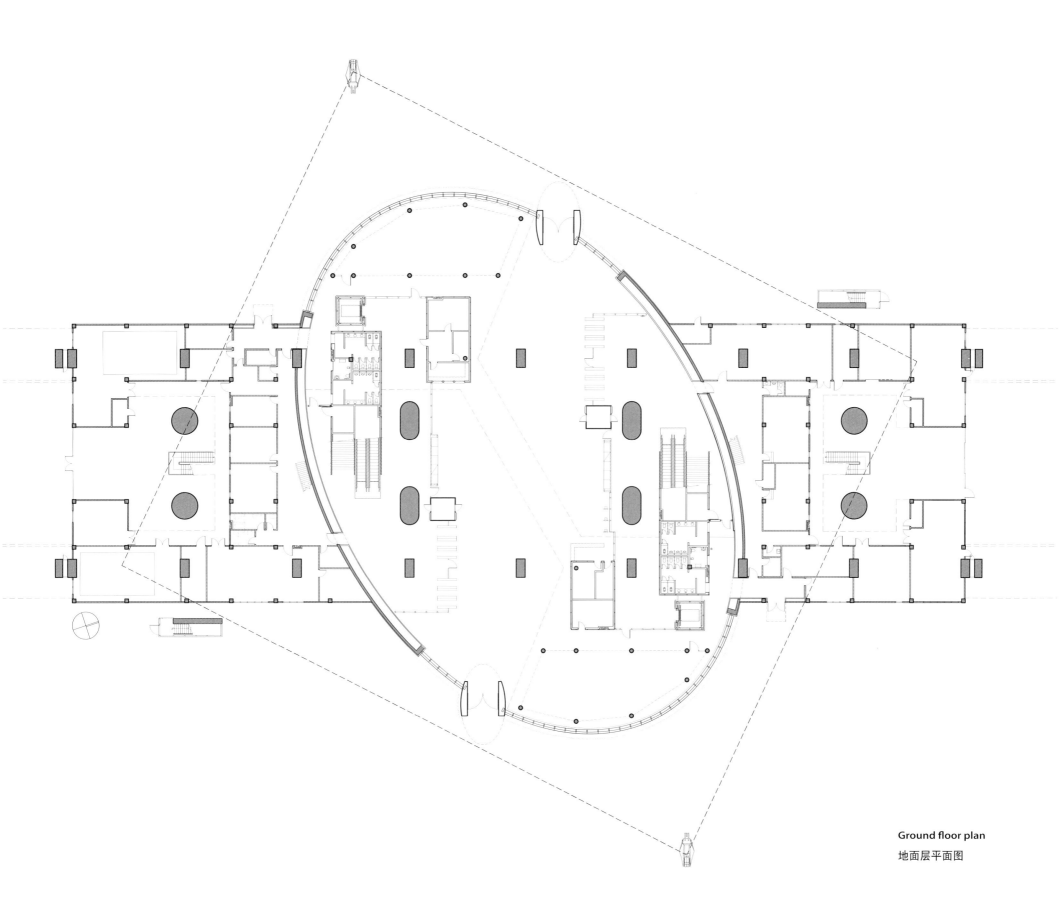

Ground floor plan
地面层平面图

新竹高速铁路站 133

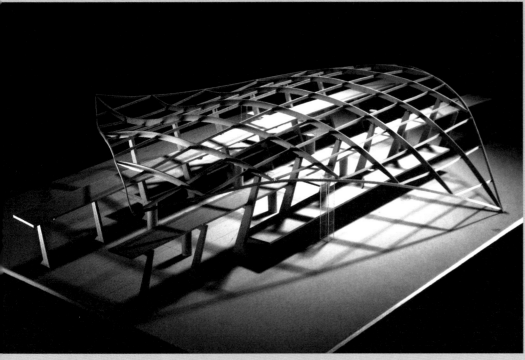

Model, station structure concept
模型，车站主结构概念

Sectional model
剖面模型

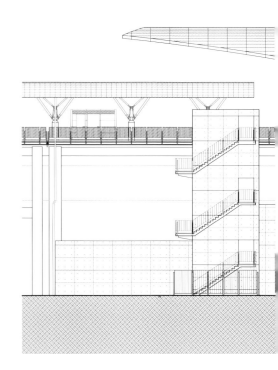

Longitudinal section
纵向剖面图

Model, aerial view
模型，鸟瞰

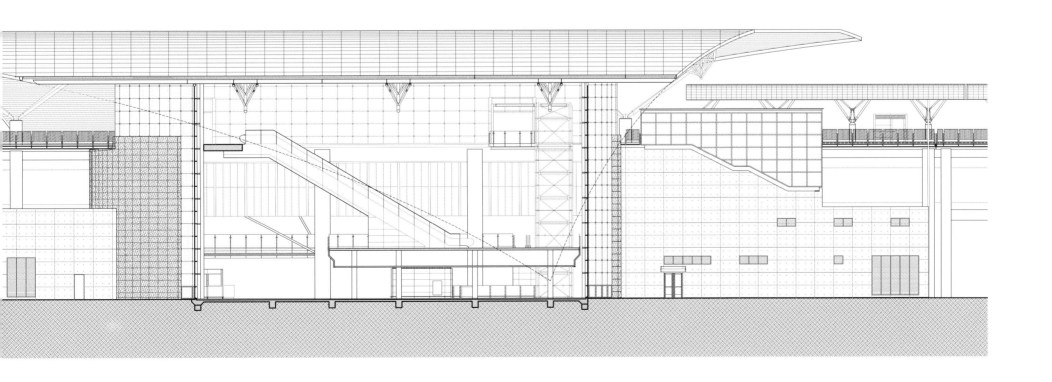

Cross section
横向剖面图

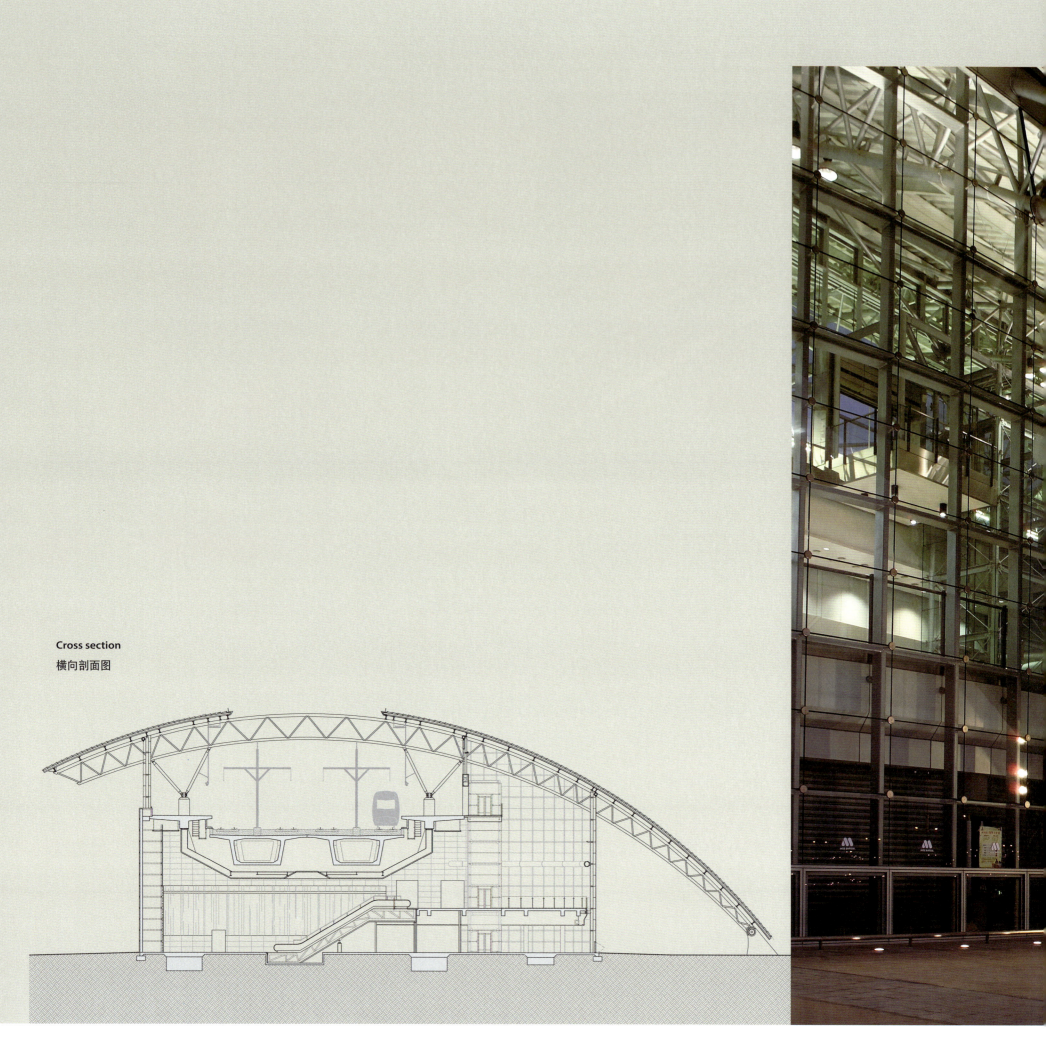

Entrance plaza
车站入口广场

新竹高速铁路站

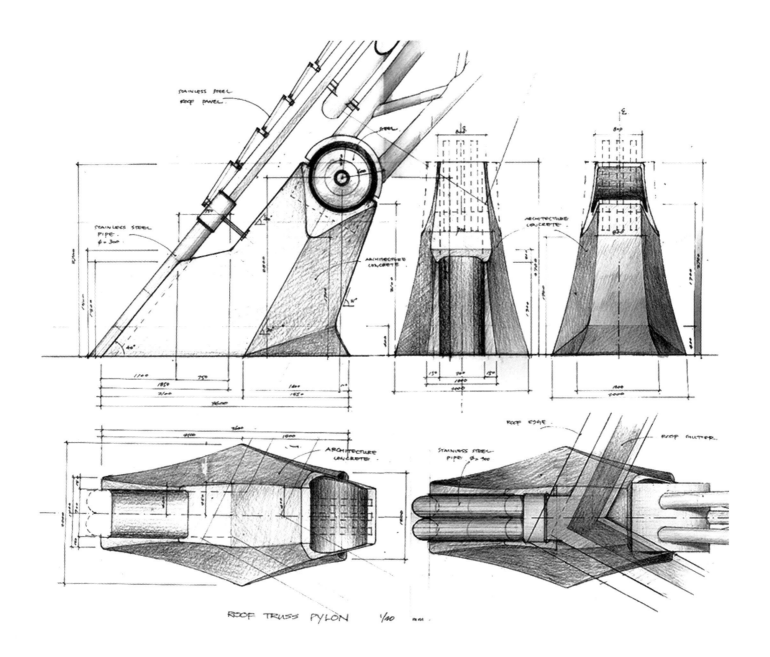

Sketch of the roof truss pylon
屋顶锚定造型柱草图

照片由 THSRC 提供

新竹高速铁路站

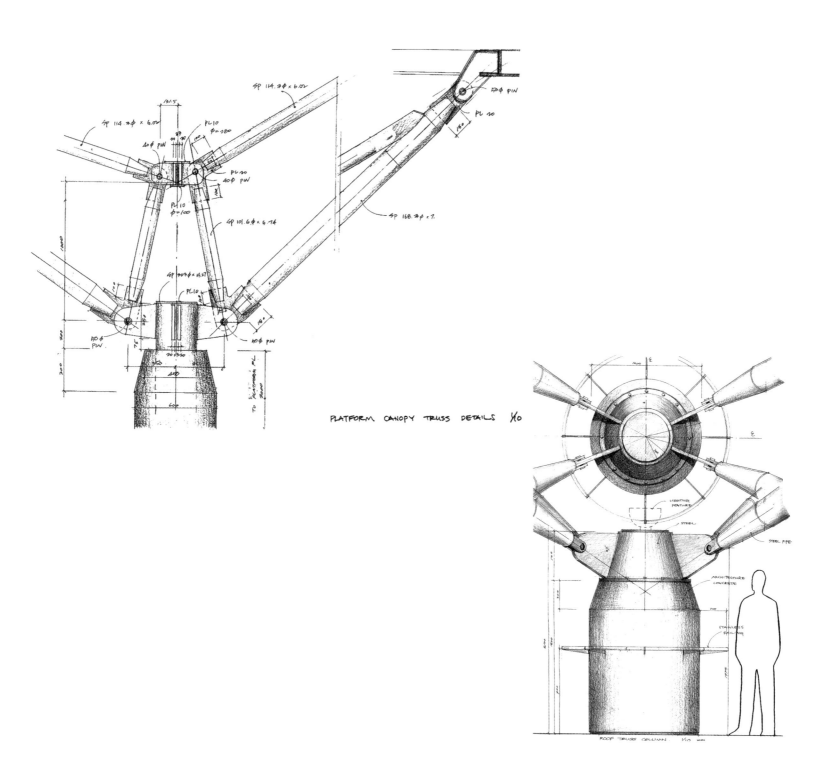

Sketch of platform canopy truss detail
月台层屋顶桁架细部草图

新竹高速铁路站

Mezzanine floor bridge
夹层空桥

新竹高速铁路站

North-bound wall art work: using glass and metal to depict "future" as its theme.
北上公共艺术墙：以玻璃及金属形塑"未来"主题。

South-bound wall art work: using folklore materials to depict "tradition" as its theme.
南下公共艺术墙：以传统民俗材料形塑"过去"主题。

新竹转运站

Hsinchu, Taiwan, China | Completion 2015
中国 台湾 新竹 | 2015 年完工

HSINCHU BUS TERMINAL

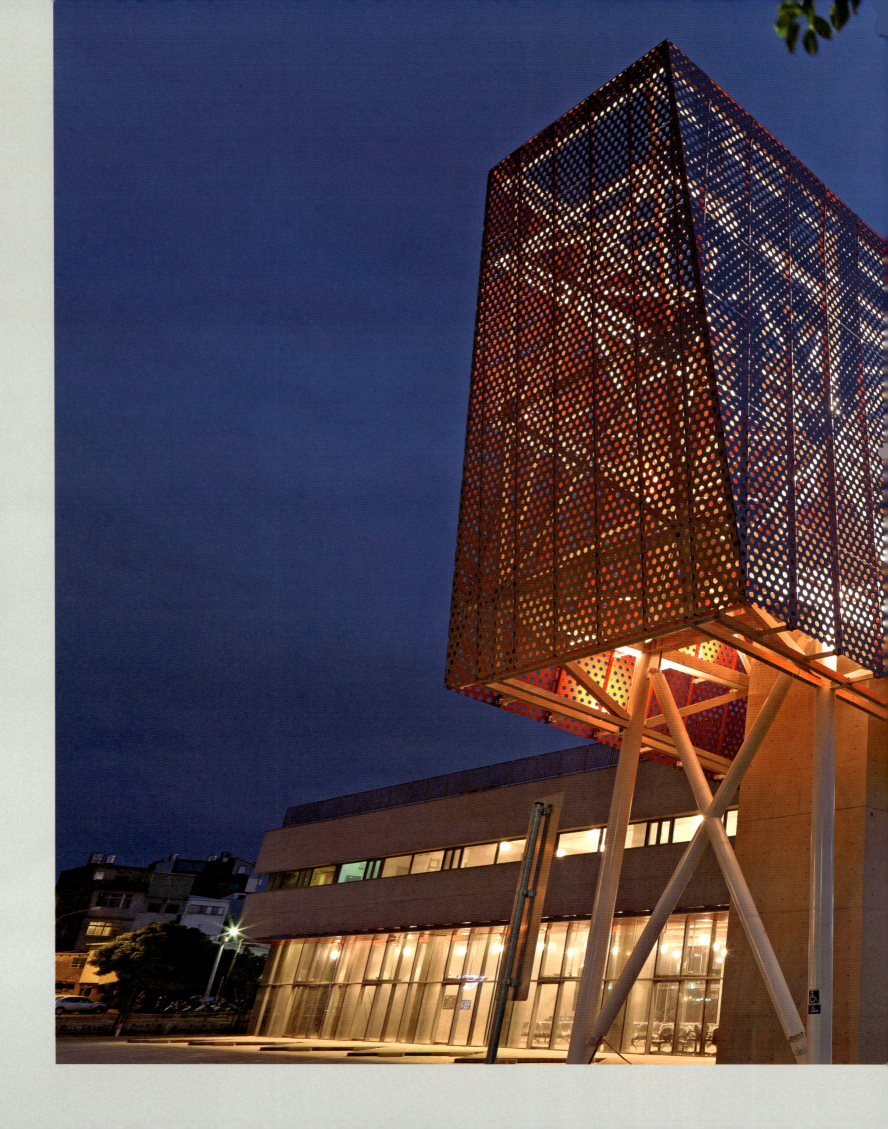

LOCATION	Hsinchu, Taiwan, China
CLIENT	Hsinchu City Government
FLOOR LEVELS	2 Floors
BUILDING STRUCTURE	Reinforced Concrete
MATERIALS	Architectural Concrete, Perforated Aluminum Panel, Glass
BUILDING USE	Station
SITE AREA	3300 ㎡
TOTAL FLOOR AREA	1462 ㎡
DESIGN INITIATIVE	2013
COMPLETION	2015

项目位址	中国 台湾 新竹
业主	新竹市政府
楼层	地上 2 层
建筑结构	钢筋混凝土
材料	清水混凝土、冲孔铝板、玻璃
用途	车站
基地面积	3300 ㎡
总楼地板面积	1462 ㎡
设计起始时间	2013
完工时间	2015

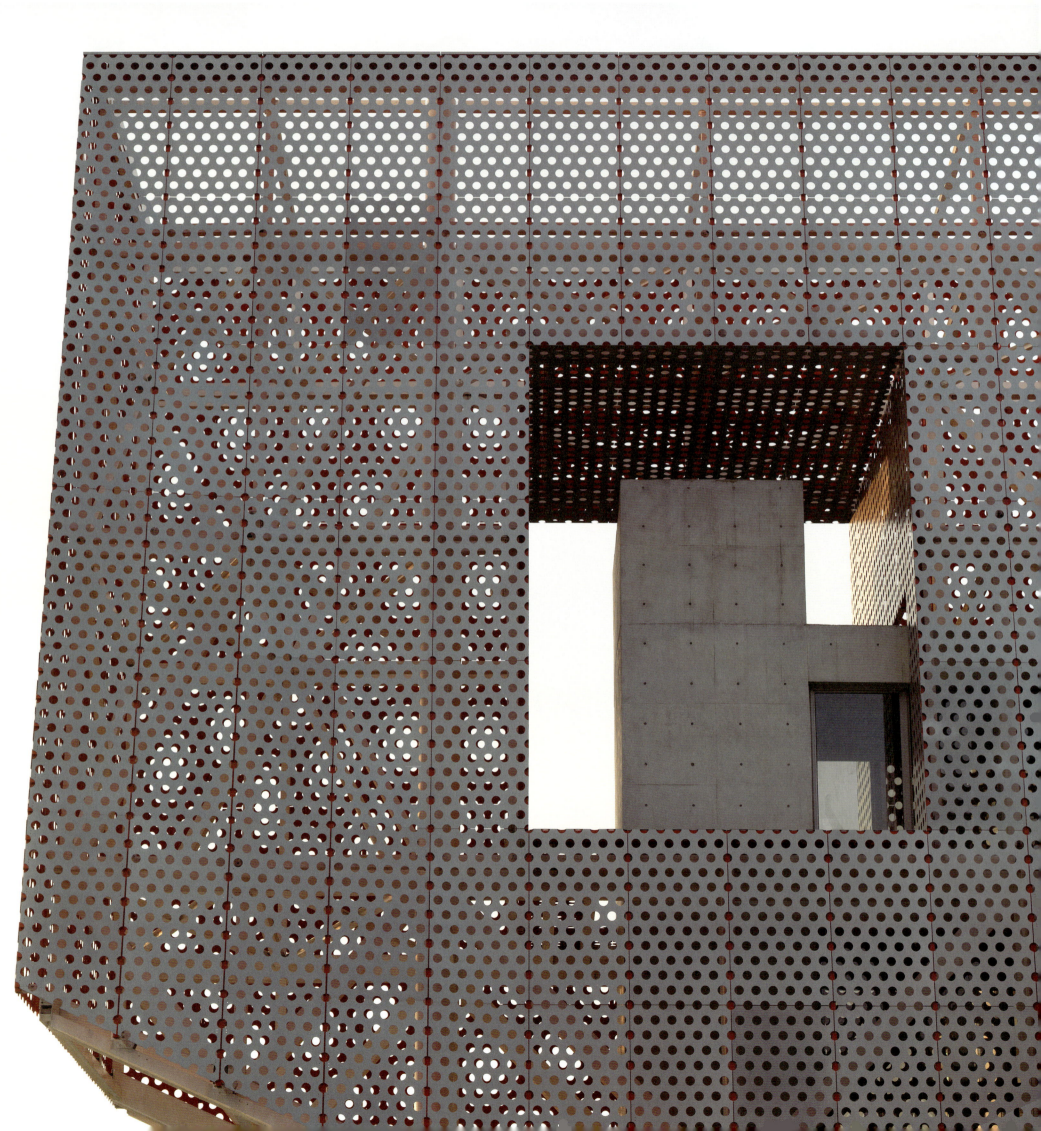

In order to relieve congestion of traffic and to enhance new urban development of the "back side" of the old Hsinchu Train Station, the City decided to move the Bus Terminal to this new location on the back of the Station. With this, the small terminal needs to achieve two purposes: to establish an efficient and safe people/bus flow on the ground level, and to create a "billboard" architecture for the public to re-orient themselves around the station area.

The two-story architectural concrete terminal building is simple and straightforward, with buses circling on the outer peripheral areas and people entering from the center, avoiding any conflicts between them. The "billboard" part of the building features lofty steel frameworks cladded with perforated aluminum panels. In the daytime, this gigantic billboard mainly shows its silver-metallic color on the outside; while in the evenings, the bright red color from inside reveals itself, giving the terminal a new urban energy and a sleek sight.

为了纾缓车站人潮与行车动线的壅塞与滞碍，并配合市区腹地发展规划，将人潮导引至后站，新竹市决定将客运转运站移至新竹车站正后方的此一基地。因此，这座小小的转运站负担了两个任务：其一，是在地面层建立一座安全有效的客运车站；其二，是创造一座"招牌"式的建筑，让市民可以对此区重新定位。

以清水混凝土建造的转运站主体相当直接明了。客运车在外围绕行，旅客自站体中央进出，有效地避免了两者可能产生的冲突。上方"招牌"造型则是以空体钢构构成，其外复冲孔铝板。在白天，此巨型招牌主要显示出金属质感的银灰色，而在夜间，冲孔铝板内部的鲜红色借由灯光露出，形成都市中新颖活泼的辨识性。

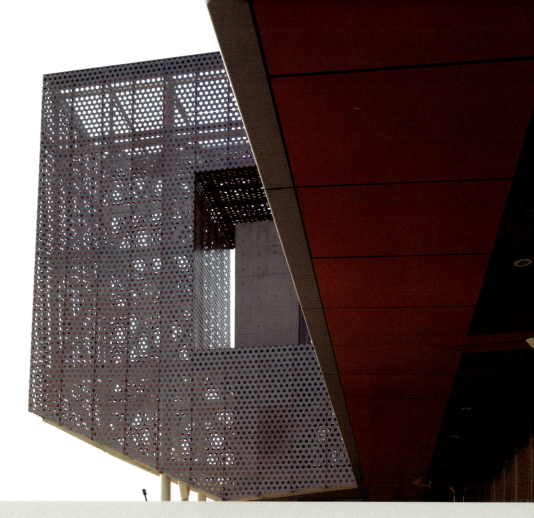

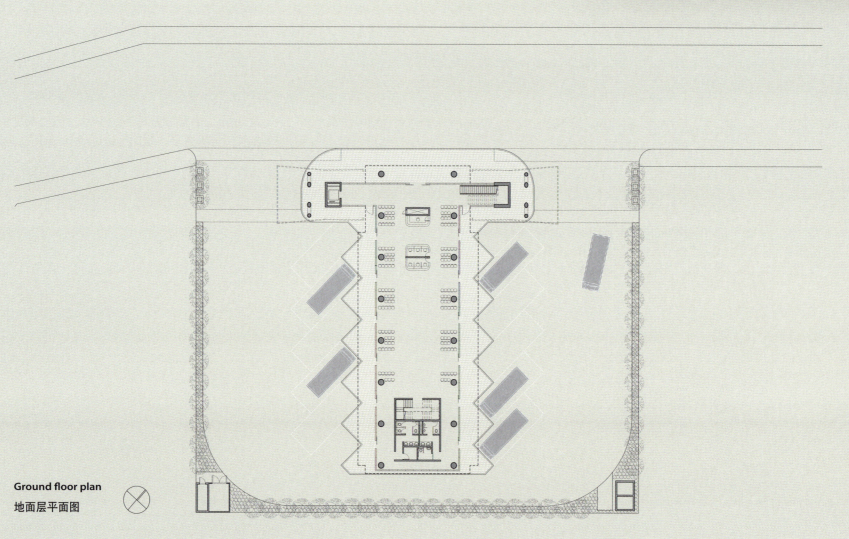

Ground floor plan
地面层平面图

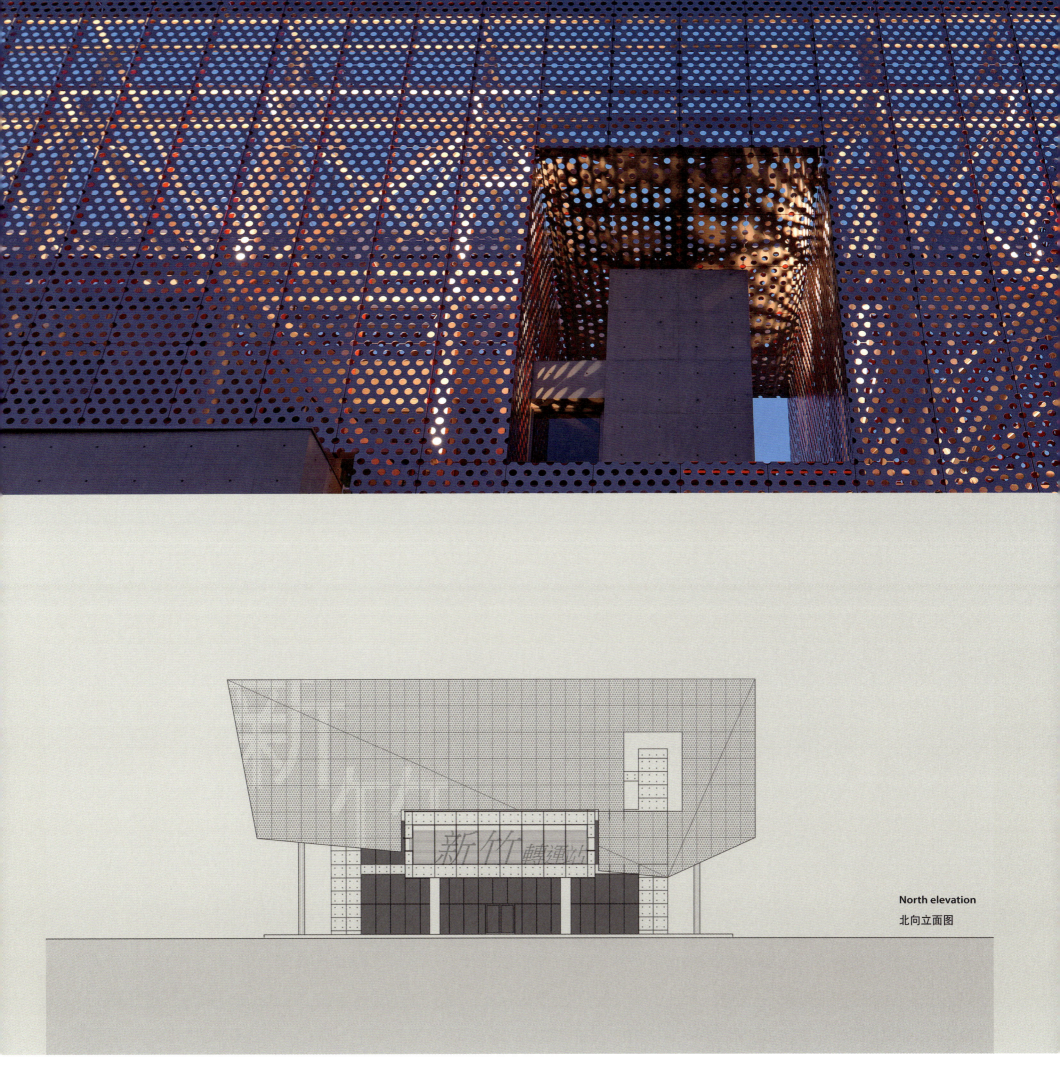

North elevation
北向立面图

新竹转运站 163

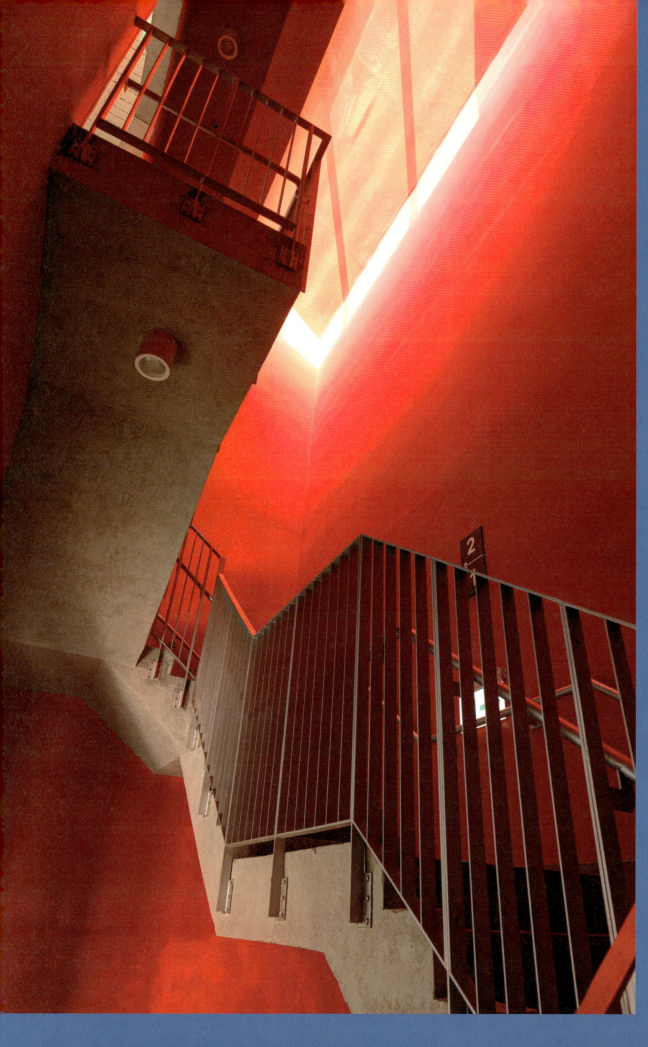

Stairway to the roof terrace
通往屋顶平台之楼梯

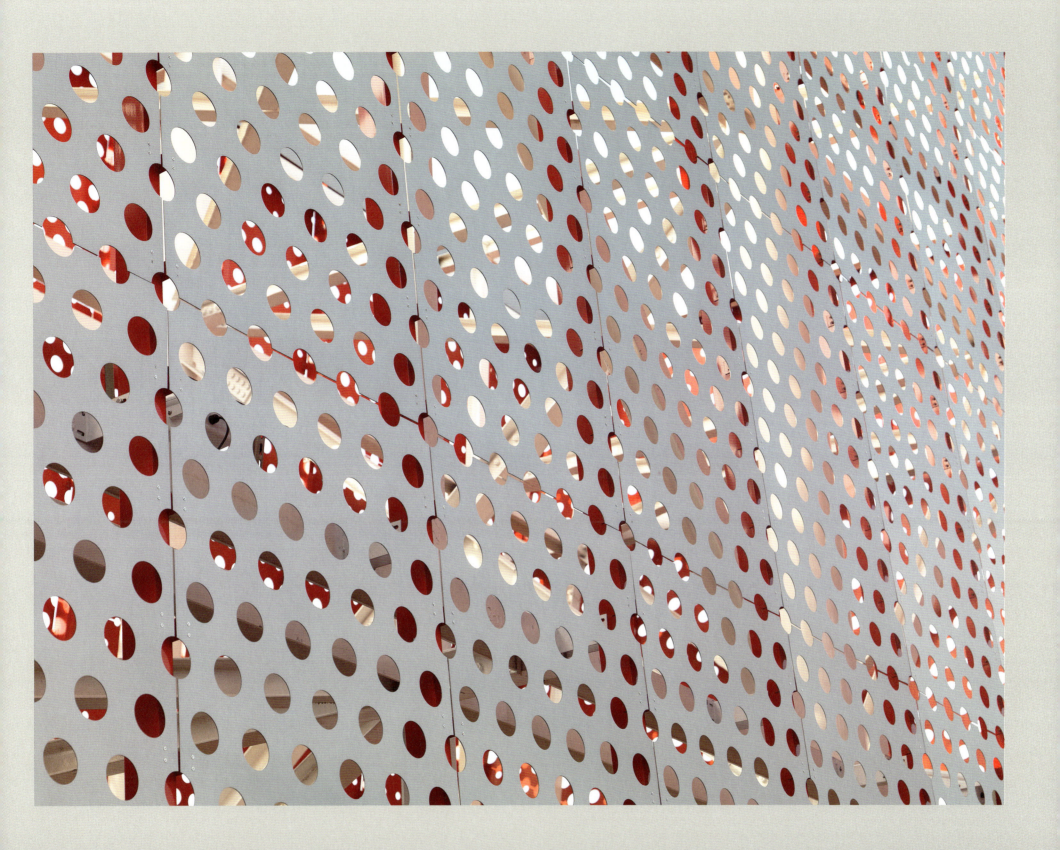

YUAN ZE UNIVERSITY LIBRARY

元智大学图书馆

Taoyuan, Taiwan, China | Completion 1998
中国 台湾 桃园 | 1998 年完工

LOCATION	Taoyuan, Taiwan, China
CLIENT	Yuan Ze University
FLOOR LEVELS	7 Floors, 1 Basement
BUILDING STRUCTURE	Pre-cast Reinforced Concrete
MATERIALS	Architectural Concrete, Extruded Aluminum Louver, Glass Unit Masonry, Clear Glass
BUILDING USE	School Facility
SITE AREA	152512 ㎡
TOTAL FLOOR AREA	84550 ㎡
DESIGN INITIATIVE	1995
COMPLETION	1998

项目位址	中国 台湾 桃园
业主	元智大学
楼层	地上 7 层、地下 1 层
建筑结构	清水混凝土预铸板
材料	清水混凝土、铝挤型百叶、玻璃砖、清玻璃
用途	学校
基地面积	152512 ㎡
总楼地板面积	84550 ㎡
设计起始时间	1995
完工时间	1998

Library at night
图书室夜景

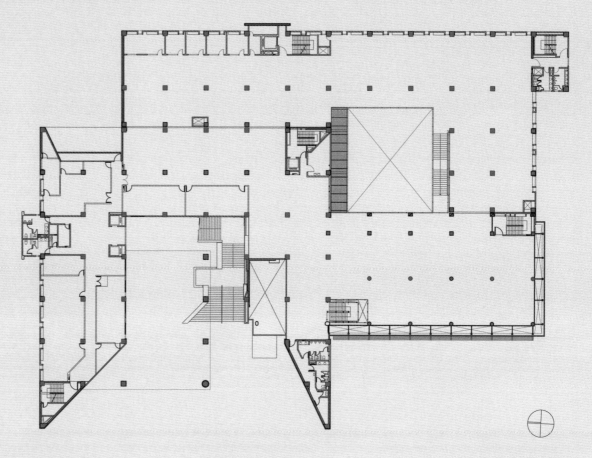

Second floor plan
二层平面图

The library is located at the end of a tree-lined boulevard, facing west, with the heavy, monumental façade, serving as an anchor building to the central plaza of the University. The building is a large portal-framed structure which houses three major functions: the L-shaped Information Science Department, a conference center in the suspended glass box, and the library section that extends out to the south. The library building is shielded by the massive concrete roof on top, and an upward passage composed of grand stairs, ramps and bridges takes people through the building to the various functions and semi-outdoor terraces at different levels facilitating impromptu exchanges and activities along the path. This Library building is not only a place for knowledge, but a focal gathering place for major activities on campus.

图书馆大楼坐落于校园林荫大道的端点，是元智大学的锚定建筑。建筑物坐东朝西，具沉厚稳重之意象，屹立于校园中央绿地广场前。图书馆大楼为大型的框架式结构，包含三项空间机能：L形空间为资讯学系中心，悬吊的玻璃方体内设置了国际会议厅，以及向南延伸的图书馆区。图书馆大楼入口由巨大混凝土屋面所庇护，中央大型楼梯、坡道及桥梁构成主要通道，引导师生至不同使用空间及半户外平台，提供多样化的活动机能与运用。本图书馆大楼不仅是富饶创意的知识殿堂，亦是校园中学生重要的交流活动的场所。

Library interior courtyard
图书馆室内中庭

Entrance detail
入口区域细部

LUODONG GOVERNMENT CENTER

罗东行政中心

Yilan, Taiwan, China | Completion 2018
中国 台湾 宜兰 | 预计 2018 年完工

LOCATION	Yilan, Taiwan, China	项目位址	中国 台湾 宜兰
CLIENT	Yilan County Government	业主	宜兰县政府
FLOOR LEVELS	6 Floors, 2 Basements	楼层	地上 6 层、地下 2 层
BUILDING STRUCTURE	Reinforced Concrete	建筑结构	钢筋混凝土
MATERIALS	Architectural Concrete, Aluminum Panel, Glass	材料	清水混凝土、铝板、玻璃
BUILDING USE	Office	用途	办公大楼
SITE AREA	12236 ㎡	基地面积	12236 ㎡
TOTAL FLOOR AREA	33044 ㎡	总楼地板面积	33044 ㎡
DESIGN INITIATIVE	2013	设计起始时间	2013
COMPLETION	2018	预计完工时间	2018

Concept sketch plan and section
平面及剖面概念图

Located at the border of Luodong Township, the Luodong Government Center is composed of a group of unusually open buildings, which will facilitate six administrative functions. Adjacent to the site is the Forestry Culture Park. The concept behind the project takes on the imagery of stacked wooden planks, with interactions between solids and voids forming along the undulating volumes.

The Center is mainly comprised of seven linear north-south facing buildings, each configured according to its functions. There are footbridges and spaces connecting the various linear building for functional purposes. The building at the center of the complex has a raised, open-air ground level that serves as a public plaza and doubles as the main route leading to the Forestry Culture Park, while exhibition spaces dotted throughout the site are available for an array of creative uses by the public. Green roofs adorn the tops of all the buildings, creating a dialogue with the nearby Forestry Culture Park. In addition to its administrative and public service functions, the center will also serve as an extension of the neighboring Forestry Culture Park, providing a diverse, user-friendly space for all visitors.

本案坐落于罗东镇入口处，由一组相当开放的建筑全体所构成。完成后，将整合宜兰的六个行政单位为"第二行政中心"，作为进出罗东的枢纽，也衔接了林业文化园区的出入口。建筑概念以木材堆叠为意象；在量体的高低起伏中，交错着虚实空间。

全区由七栋南北向的长型建筑量体所组成，各依不同的行政单位配置。位于中间的量体地面层抬高留白，成为民众的活动广场及连接林业文化园区的主要路径，其中穿插的创意展示空间则提供给民众多元运用。各栋楼层除了垂直动线外，亦有水平的楼梯、户外空间与空桥联系；楼顶的绿化植栽将"以林为邻"的概念呼应林业文化园区。本中心除了具备服务民众的行政功能之外，也作为附近林业园区休闲遊憩的据点的延伸，成为多元化的空间。

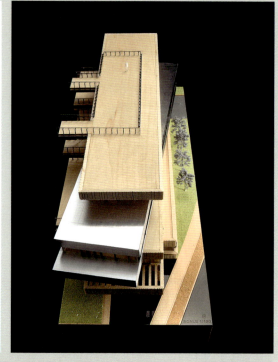

Model, entrance to the administration center
模型，行政中心入口

Concept: a stack of wood plank at Luodong Forestry Culture Park
概念示意：堆迭于罗东林业文化园区的木材

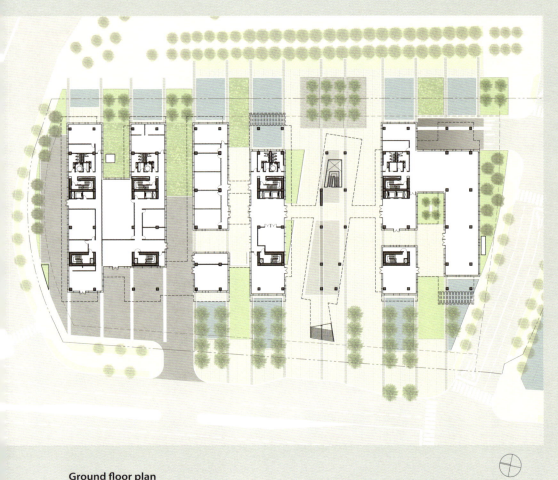

Ground floor plan
地面层平面图

Second floor plan
二层平面图

186 LUODONG GOVERNMENT CENTER

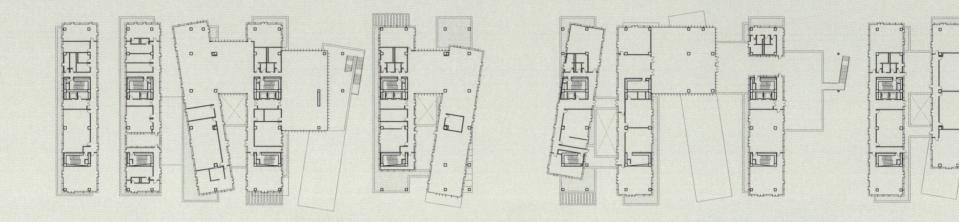

Third floor plan
三层平面图

Fourth floor plan
四层平面图

West elevation
西向立面图

Longitudinal sectional elevation
纵向剖立面图

Cross sectional elevations
横向剖立面图

罗东行政中心　189

CHRONOLOGICAL LIST OF SELECTED PROJECTS

精选作品年表

Grand Formosa Regent
Taipei, Taiwan, China. Completed
丽晶酒店
中国 台湾 台北，完工

Weigo Elementary School
Taipei, Taiwan, China, Completed
薇阁小学
中国 台湾 台北，完工

Yuan-Ze University College of Engineering
Taoyuan, Taiwan, China, Completed
元智大学工学院
中国 台湾 桃园，完工

Jen Ai Tower
Taipei, Taiwan, China, Completed 1989
鸿禧仁爱住宅大楼
中国 台湾 台北，1989年完工

The Linden Hotel
Kaohsiung, Taiwan, China, Completed 1994
高雄雷园饭店
中国 台湾 高雄，1994年完工

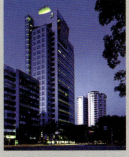

Fubon Banking Center
Taipei, Taiwan, China, Completed 1995
富邦仁爱金融中心
中国 台湾 台北，1995年完工

Collaborating Architects: SOM
合作建筑师：SOM

Sheng Yang Banking Center
Taipei, Taiwan, China, Completed 1995
升阳国际金融中心
中国 台湾 台北，1995年完工

Fubon Kaohsiung Office Building
Kaohsiung, Taiwan, China, Completed 1997
富邦建设高雄民族路办公大楼
中国 台湾 高雄，1997年完工

Tung Hwa Univeristy Faculty Club
Hualien, Taiwan, China, Completed 1995
东华大学教职员餐厅
中国 台湾 花莲，1995年完工

225 Tung Hwa South Road
Taipei, Taiwan, China, Completed 1996
元大建设敦化南路225号
中国 台湾 台北，1996年完工

Fubon Life Insurance Headquarters
Taipei, Taiwan, China, Completed 1999
富邦人寿总部大楼
中国 台湾 台北，1999年完工

Continental Engineering Corporation Headquarters
Taipei, Taiwan, China,Completed 1999
汉德大楼
中国 台湾 台北，1999年完工

Yuan Ze University Library
Taoyuan, Taiwan, China, Completed 1998
元智大学图书馆
中国 台湾 桃园，1998年完工

Luminary Buddhist Center
Taichung, Taiwan, China, Completed 1998
养慧学苑
中国 台湾 台中，1998年完工

Yageo Corp. Kaohsiung Plant II
Kaohsiung, Taiwan, China, Completed 1997
国巨电子高雄第二厂房
中国 台湾 高雄，1997年完工

TNNUA, College of Sound and Image Arts
Tainan, Taiwan, China, Completed 1998
台南艺术大学音像艺术馆
中国 台湾 台南，1998年完工

P Residence
Taipei, Taiwan, China, Completed 1999
P宅
中国 台湾 台北，1999年完工

Taiwan GSM Headquarters
Taipei, Taiwan, China, Completed 2001
台湾大哥大企业总部
中国 台湾 台北，2001年完工

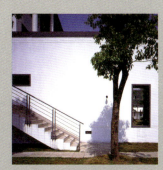

Yageo Suzhou Industrial Complex
Suzhou, China, Completed 1998
国巨电子苏州厂房
中国 苏州，1998年完工

Hsinchu High Speed Rail Station
Hsinchu, Taiwan, China, Completed 2006
新竹高速铁路站
中国 台湾 新竹，2006年完工

Compal Headquarters
Taipei, Taiwan, China, Completed 1999
仁宝企业总部大楼
中国 台湾 台北，1999年完工

Lite-On Headquarters
Taipei, Taiwan, China, Completed 2002
光宝科技大楼
中国 台湾 台北，2002年完工

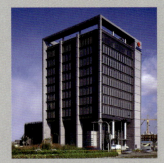

MITAC Headquarters
Taipei, Taiwan, China, Completed 2000
神通电脑企业总部大楼
中国 台湾 台北，2000年完工

Shin Min High School Music Center
Taichung, Taiwan, China, Completed 1999
新民音乐中心
中国 台湾 台中，1999年完工

Premier Headquarters
Taipei, Taiwan, China, Completed 2001
普立尔科技大楼
中国 台湾 台北，2001年完工

SET TV Headquarters
Taipei, Taiwan, China, Completed 2001
三立电视大楼
中国 台湾 台北，2001年完工

Shih Chien University College of Design
Taipei, Taiwan, China, Completed 2003
实践大学设计学院
中国 台湾 台北，2003年完工

Yuan-Ze University Administration and Academy Building
Taoyuan, Taiwan, China, Completed 2002
元智大学行政教学大楼
中国 台湾 桃园，2002年完工

Lanyang Museum
Yilan, Taiwan, China, Completed 2010
兰阳博物馆
中国 台湾 宜兰，2010年完工

The Cypress Court
Taipei, Taiwan, China, Completed 2010
元大栢悦
中国 台湾 台北，2010年完工

Fu Shin School
Taipei, Taiwan, China, Completed 2007
复兴中小学
中国 台湾 台北，2007年完工

WK Technology Headquarter
Taipei, Taiwan, China, Completed 2008
晋禾大楼
中国 台湾 台北，2008年完工

Peitou Catholic Church
Taipei, Taiwan, China, Competition
北投天主堂
中国 台湾 台北，竞图

Quanta Research & Development Center
Taoyuan, Taiwan, China, Completed 2004
广达研发中心
中国 台湾 桃园，2004年完工

Tomihiro Art Museum
Gunma, Japan, Competition
富弘美术馆竞图
日本 群马，竞图

Guang Ren Elementary School
Taipei, Taiwan, China, Completed 2005
天主教光仁小学
中国 台湾 台北，2005年完工

Hua Hsing High School
Taipei, Taiwan, China, Completed 2006
华兴中学
中国 台湾 台北，2006年完工

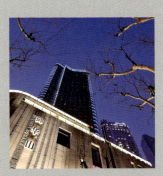

Huashan House
Shanghai, China, Completed 2008
上海御华山
中国 上海，2008年完工

Want Want Group Headquarters
Shanghai, China, Completed 2008
旺旺总部
中国 上海，2008年完工

Tongshan Residential Tower
Taipei, Taiwan, China, Completed 2007
铜山街集合住宅
中国 台湾 台北，2007年完工

Yuanta Financial Tower
Taipei, Taiwan, China, Completed 2007
元大金控大樓
中国 台湾 台北，2007年完工
Architect of Record: TMA Architects & Associates
执行建筑师：三门建筑师事务所

The Drape House
Nanjing, China, Design project
佛手湖墅
中国 南京，设计方案

Shih Chien University Gymnasium and Library
Taipei, Taiwan, China, Completed 2009
实践大学体育馆+图书馆
中国 台湾 台北，2009年完工

Chiao Tung University Guest House
Hisnchu, Taiwan, China, Completed 2008
交通大学招待所
中国 台湾 新竹，2008年完工

Meridian Hotel and Office Complex
Taipei, Taiwan, China, Completed 2010
華舍艾美酒店商务中心
中国 台湾 台北，2010年完工

Xinyi Shopping Center
Taipei, Taiwan, China, to be Completed in 2018
信义购物中心
中国 台湾 台北，预计2018年完工

Songshan Line Metro, Nanjing S. Rd. & Taipei Gymnasium Station
Taipei, Taiwan, China, Completed 2014
捷运松山线 南京东路站及市立体育场站
中国 台湾 台北，2014年完工

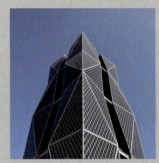

China Steel Corporation Headquarters
Kaoshung, Taiwan, China, Completed 2012
中钢企业总部大楼
中国 台湾 高雄，2012年完工

Fo Guang Shan Monastery
Vienna, Austria, Completed 2010
佛光山维也纳禅净中心
奥地利 维也纳，2010年完工

Fo Guang Shan Monastery
North Carolina, U.S.A., Completed 2008
佛光山北卡萝萊纳禅净中心
美国 北卡萝萊，2008年完工

Fo Guang Shan Monastery
Paris, France, Completed 2012
巴黎佛光山
法国 巴黎，2012年完工

Lite-On IT Research & Development Center
Hisnchu, Taiwan, China, Completed 2006
建兴电子研发中心
中国 台湾 新竹，2006年完工

Merck Liquid Crystal Center
Taoyuan, Taiwan, China, Completed 2005
默克光电液晶生产暨研发中心
中国 台湾 桃园，2005年完工

Far Eastern Telecom Park
New Taipei, Taiwan, China, Completed 2010
远东通讯园区研发大楼
中国 台湾 新北，2010年完工

Palace Museum Southern Branch
Chiayi, Taiwan, China
2004 International Competition
故宫南院-2004国际竞图
中国 台湾 嘉义，竞图

Wushih Beach Pavilion
Yilan, Taiwan, China, Completed 2007
乌石海水浴场游客中心
中国 台湾 宜兰，2007年完工

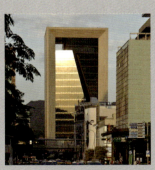

Kelti Center
Taipei, Taiwan, China, Completed 2009
克缇大楼
中国 台湾 台北，2009年完工

Fubon 777 Residential Tower
Taipei, Taiwan, China, Completed 2011
富邦天母777集合住宅
中国 台湾 台北，2011年完工

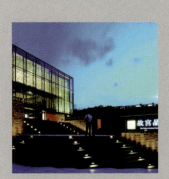

Silks Palace, Palace Museum (Taipei)
Taipei, Taiwan, China, Completed 2008
故宫晶华
中国 台湾 台北，2008年完工

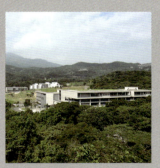

Dharma Drum Institute of Liberal Arts
New Taipei, Taiwan, China, Completed 2015
法鼓文理学院
中国 台湾 新北，2015年完工

TIAA MRT System
Taoyuan, Taiwan, China,
to be completed in 2015
机场捷运联外系统
中国 台湾 桃园，预计2015年完工

Shanghai Commercial & Savings Bank Residential Tower
Taipei, Taiwan, China, Completed 2010
上海商银仁爱住宅案
中国 台湾 台北，2010年完工

Water-Moon Monastery
Taipei, Taiwan, China, Completed 2012
水月道场
中国 台湾 台北，2012年完工

Ching Shin School
Taipei, Taiwan, China, Completed 2010
静心中小学
中国 台湾 台北，2010年完工

Weiwuying Perfroming Arts Center
Kaoshung, Taiwan, China, Competition
卫武营艺文中心
中国 台湾 高雄，竞图

Design Collaboration: Chien Architects & Associates/
Stonehenge Architects International
合作建筑师：竹间联合建筑师事务所／大硯国际建筑师事务所

Chiaokuo Residential Tower
Taipei, Taiwan, China, Completed 2010
侨果复兴铜山街
中国 台湾 台北，2010年完工

Kunming University of Science and Technology Library, Auditorium & Classroom Buildings
Yunnan, China, Completed 2009
昆明理工大学
中国 云南，2009年完工

Far Eastern Banqiao Skyscraper
New Taipei, Taiwan, China, Completed 2014
远东百货板桥办公大楼
中国 台湾 新北，2014年完工

Master Plan & Department Store Building Design by:
Kisho Kurokawa Architects & Associates
整体规划及百货建筑设计：黑川纪章建筑都市设计事务所

Taiwan NEXT-GENE Ao-Di Project
New Taipei, Taiwan, China, Design project
澳底年澳底计划
中国 台湾 新北，设计案

Changhua High Speed Rail Station
Changhua, Taiwan, China,
to be Completed in 2015
彰化高速铁路站
中国 台湾 彰化，预计2015年完工

Yunlin High Speed Rail Station
Yunlin, Taiwan, China,
to be Completed in 2015
云林高速铁路站
中国 台湾 云林，预计2015年完工

Cathay Landmark
Taipei, Taiwan, China, to be Completed in 2015
国泰置地广场
中国 台湾 台北，预计2015年完工

Executive Architect: Hcch & Associates Architects
Planners & Engineers
执行建筑师：三大联合建筑师事务所

HTC Taipei Headquarters
New Taipei, Taiwan, China, Completed 2012
宏达电台北总部大楼
中国 台湾 新北，2012年完工

Hua Nan Bank Headquarters
Taipei, Taiwan, China, Completed 2014
华南银行总部
中国 台湾 台北，2014年完工

LE Office Tower
Taipei, Taiwan, China, Design project
光世代办公大楼
中国 台湾 台北，设计案

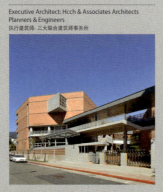

Humble House
Taipei, Taiwan, China, Completed 2013
寒舍艾丽酒店
中国 台湾 台北，2013年完工

TAS Facilities Development Project
Taipei, Taiwan, China, Completed 2012
美国学校增建工程
中国 台湾 台北，2012年完工

Quemoy University Student Center
Kinmen, Taiwan, China, Completed 2014
金门大学学生活动中心
中国 台湾 金门，2014年完工

Pingtung Performing Arts Center
Pingtung, Taiwan, China, Completed 2015
屏东演艺厅
中国 台湾 屏东，2015年完工

United Daily News Group Complex
Taipei, Taiwan, China, to be Completed in 2017
联合报办公及住宅大楼
中国 台湾 台北，预计2017年完工

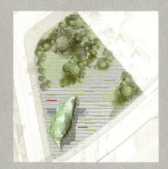
Taiwan Pavilion of Shanghai World Expo
Shanghai, China, Competition
上海世博台湾馆竞图
中国 上海，竞图

Taiwan Xiqu Center
Taipei, Taiwan, China, to be Completed in 2015
台湾戏曲中心
中国 台湾 台北，预计2015年完工

Museum of Prehistory
Tainan, Taiwan, China, to be Completed in 2016
史前博物馆
中国 台湾 台南，预计2016年完工

Fengqiyong Chinese Art Exhibition Hall
Jiangsu, China, Completed 2011
冯其庸学术馆
中国 江苏，2011年完工

Eslite Suzhou
Jiangsu, China, to be Completed in 2015
苏州诚品
中国 江苏，预计2015年完工

Far Eastone Telecommunications Headquarters
New Taipei, Taiwan, China, to be Completed in 2017
远传电信总部
中国 台湾 新北，预计2017年完工

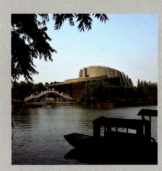
Wuzhen Theater
Zhejiang, China, Completed 2013
乌镇剧院
中国 浙江，2013年完工

Palace Museum Southern Branch
Chiayi, Taiwan, China, to be Completed in 2015
故宫南院
中国 台湾 嘉义，预计2015年完工

Hsinchu Bus Terminal
Hsinchu, Taiwan, China, Completed 2015
新竹转运站
中国 台湾 新竹，2015年完工

Huangshan Exhibition Center
Anhui, China, to be Completed in 2015
黄山城市展示馆
中国 安徽，预计2015年完工

Simatai Gubei Hotel
Beijing, China, to be Completed in 2015
司马台古北大酒店
中国 北京，预计2015年完工

Luodong Government Center
Yilan, Taiwan, China, to be Completed in 2018
罗东行政中心
中国 台湾 宜兰，预计2018年完工

NTU Cosmology Center
Taipei, Taiwan, China, to be Completed in 2017
台大宇宙学馆
中国 台湾 台北，预计2017年完工

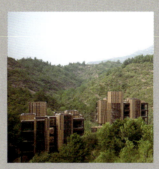
Hangu Villa
Beijing, China, Completed 2015
函谷山庄
中国 北京，2015年完工

Sanlitun Beijing
Beijing, China, Design project
北京三里屯
中国 北京，设计案

Hotel Indigo Taipei
Taipei, Taiwan, China, to be Completed in 2017
台北英迪格酒店
中国 台湾 台北，预计2017年完工

Wutaishan Retreat Center
Shanxi, China, to be Completed in 2017
五台山禅修中心
中国 山西，预计2017年完工

New Taipei Museum of Art
New Taipei, Taiwan, China, to be Completed in 2019
新北美术馆
中国 台湾 新北，预计2019年完工

Baohua Mountain Retreat Center
Jiangsu, China, to be Completed in 2015
南京宝华山
中国 江苏，预计2015年完工

FOXCONN Shanghai
Shanghai, China, to be Completed in 2017
富士康上海
中国 上海，预计2017年完工

COLLABORATORS
合作团队

3DC Concept Designers Ltd.		Lerch, Bates & Associates Inc.	
Alfredo Arribas Arquitectos Asociados		Leuchte Lighting Design Co., Ltd.	洛西特灯光设计顾问有限公司
ALT Cladding Design Inc.	美商艾勒泰工程顾问股份有限公司台湾分公司	Li Jeng Wei Structure Engineers office	栗正玮结构技师事务所
Ans Consulting Engineers Ltd.	安美设计顾问有限公司	Lincolne Scott Group	
Arup	奥雅纳	Magnificence Interiars Inc.	禾安室内装修设计工程股份有限公司
Belt Collins & Associates, Hong Kong	贝尔高林国际(香港)有限公司	Majestic Engineering Consultants, Inc.	明智工程顾问有限公司
Bensley Design Group International		Majetech Electrical Engineers Office	铭竟电机技师事务所
Bovis Lend Lease Corporation	美商劭尔工程顾问有限公司台湾分公司	Mandartech Interiors Inc.	元崇设计工程股份有限公司
Carol R. Johnson Associates Inc. Boston, U.S.A.	美国卡罗·约翰逊景观规划设计师事务所	Marco Façade Studio	马可卢设计技术有限公司
CERMAK PETERKA PETERSEN, U.S.A		Ming Sheng Engineering Co., Ltd.	茗生工程股份有限公司
Chin Tai Feng Consulting Inc.	锦泰丰工程有限公司	Moh and Associates, Inc.	亚新工程顾问有限公司
Chiu Hui Huang Structure Engineers office	邱辉煌结构技师事务所	Mori Building Co., Ltd.	森大厦株式会社
Chroma 33 Architectural Lighting Design	大公照明设计顾问有限公司	Ocean Construction International Corporation	远洋国际建设有限公司
China Academy of Art	中国美院	Office for Metropolitan Architecture (OMA)	大都会建筑事务所(OMA)
Chuang Wei Structural Engineering Inc.	创纬工程顾问有限公司	Parsons Brinckerhoff	美商柏诚
CNHW Planning & Design Consultants	禾拓规划设计顾问有限公司	Pei Cobb Freed & Partners Architects	
Connell Wagner International Pty Ltd. Australia		Pelli Clarke Pelli Architects	佩里·克拉克·佩里建筑师事务所
Continental Engineering Corp.	大陆工程股份有限公司	Performance, Arts, Technology, Design Consultant Inc.	乙太设计顾问有限公司
Continental Engineering Consultants, Inc.	大陆设备工程顾问有限公司	PLACEMEDIA, Landscape Architects Collaborative	
CO-YOUNG ENGINEERING CONSULTANTS, Inc.	协展冷冻空调技师事务所	Porun Consulting Engineering, Inc.	保伦工程顾问有限公司
Creative Solution Integration LTD.	行式管理顾问股份有限公司	Renzo Piano Building Workshop	伦佐·皮亚诺建筑工作室
Curtain Wall Design & Consulting, Dallas, TX, U.S.A.	美商希迪西(CDC)	Rowan Williams Davies & Irwin Inc.	
Dachung Survey Ltd.	大众测绘有限公司	RTKL International Ltd.	亚图建筑设计咨询有限公司
David Consulting Inc.	筑友工程顾问有限公司	San Luis Sustainability Group	
Dillingham Associates Landscape Architects	地灵国际工程顾问有限公司	Segreene Design and Consulting	澄毓设计顾问有限公司
DMJM International		Shanghai Xian Dai Architectural Design (Group) Co., Ltd.	上海现代建筑设计(集团)有限公司
East China Architectural Design & Research Institute Co., Ltd.	华东建筑设计研究院有限公司	Shanghai Institute of Architectural Design & Research	上海建筑设计研究院有限公司
Easygogo Technology Inc.	行易网科技股份有限公司	S. H. Chen Partners & Associates Structure Engineering	陈水心土木结构技师事务所
EDAW Urban Design Limited		S. H. Chiang Structural Eng.	江世雄结构技师事务所
ennead architects		Shen Milsom & Wilke Limited	声美华有限公司
EDS International Inc.	境群国际规划设计顾问股份有限公司	Sincerity Engineering Consultants, Ltd.	诚驿工程顾问有限公司
Elite Consultants Engineering Associates	泉成机电工程顾问有限公司	Sino Geotechnology., Inc.	富国技术工程服务有限公司
Envision Engineering Consultants	筑远工程顾问有限公司	Skidmore, Owings & Merrill LLP.	SOM建筑设计事务所
Evergreen Consulting Engineering, Inc.	永峻工程顾问股份有限公司	SLA	Studio • Land	
Federal Engineering Consultants, Ltd	联邦工程顾问有限公司	Structural Design Group Co., Ltd	株式会社构造设计集团
FMS Partners in Architectural Lighting Design	美国FMS照明设计	Sun Te-Hung Associates	孙德鸿建筑师事务所
Fountainhead Design Studio (Shanghai) Co., Ltd.	领泉室内设计顾问（上海）有限公司	Supertech Consultants International	超伟工程顾问有限公司
Gensler	金斯勒	Suzhou Institute of Architectural Design & Research	苏州市建筑设计研究院
gmp Architekten von Gerkan, Marg und Partner	gmp冯·格康，玛格及合伙人建筑师事务所	Taiwan Fire Safety Consulting, Ltd.	台湾建筑与都市防灾顾问有限公司
Grand Ages Surveying Consultants Company Limited	大时代测量顾问有限公司	Taiwan Green Architecture	台湾绿建筑
Greentown Oriental Architects (GOA)	北京绿城东方建筑设计有限公司	Taiwan University of Science and Technology	台湾科技大学
H&K Associates	康普工程顾问股份有限公司	Tadao ando Architects & Associates. Osaka, Japan	安藤忠雄建筑研究所
Han Ming Landscape Architects Associates	汉明景观有限公司	Takano Landscape Planning Co., Ltd.	日商日亚高野景观规划(股)公司台湾分公司
Handar Engineering & Construction Inc.	汉达工程顾问有限公司	The Jerde Partnership, International, Inc.	捷得建筑师事务所
Heng Kai Engineering Consultants, Inc.	恒开工程顾问有限公司	The SWA Group	SWA景观设计公司
Huai Te Engineering Consultants, Inc.	怀德技术顾问有限公司	Theatre Projects Consultants Ltd.	剧院项目咨询有限公司
Huang Wan Fu Civil Engineers office	黄万福土木技师事务所	THI Consultants Inc.	鼎汉国际工程顾问股份有限公司
Hyder Consulting Limited	安诚集团	Tino Kwan Lighting Consultants	关永权照明设计公司
Innerscapes Designs	内海设计有限公司	Tongji University Tongda Construction Consultant	同济大学同大建筑工程咨询事务所
I. S. Lin and Associates Consulting Engineers	林伸环控设计有限公司	TOPO Design Group	
Jauhung Consulting Engineers	昭宏工程顾问有限公司	T. S Wang Archietect & Associates	王德生建筑师事务所
Jellys Technology Inc.	捷力士科技有限公司	T. Y. Lin Taiwan Consulting Engineers, Inc.	林同棪工程顾问(股)公司
Jia-mao Construction Co., Ltd.	家茂营造工程有限公司	Tzerong Engineering Co., Ltd.	泽荣工程有限公司
Jun House Square Consulting Engineering, Inc.	俊屋坊工程顾问有限公司	UD Architectural Design Institute	现代都市设计院 项目设计十一部
Kai-Chu Engineering Consultant, Inc.	凯巨工程顾问有限公司	Unitech Engineering Inc.	光宇工程顾问有限公司
Kaichuan Planning Design Co., Ltd.	开创规划设计股份有限公司	WeBim Services Co., Ltd.	卫武信息股份有限公司
Kaishun Survey Company	凯顺测量有限公司	Weiskopf & Pickworth Structural Engineers	
Rottet Studio Architecture and Design		Wilmotte & Associes SA / Nelson Wilmotte Architectes	
Kenzo Tange Associates Urbanists-Architects	丹下健三都市设计研究所	WINDTECH Consultants Pty Ltd.	
KIGHTON Facade Consultants	上海凯腾幕墙设计咨询有限公司	Xian Dai Architectural Design	上海现代建筑装饰环境设计研究院有限公司
King - Le Chang & Associates	杰联国际工程顾问有限公司	XU-ACOUSTIQUE	徐氏声学艺术顾问
Kisho Kurokawa Architect & Associates	黑川纪章建筑都市设计事务所	Yuantai Consultant Engineer	元泰工程顾问有限公司
Kuang Yu Engineering Consultants, Inc.	光友工程顾问有限公司	Zaha Hadid Architects	扎哈·哈迪德建筑师
Kun-Tai Consultants & Associates	坤泰电机技师事务所	ZEB-Technology Pte. Ltd.	
Lab-7 International Design Co., Ltd.	七观国际有限公司		

图书在版编目（CIP）数据

Communal Forums 聚 / 姚仁喜著. — 沈阳：辽宁科学技术出版社，2015.8
 ISBN 978-7-5381-9372-5

Ⅰ.①C… Ⅱ.①姚… Ⅲ.①建筑设计－作品集－中国－现代 Ⅳ.①TU206

中国版本图书馆 CIP 数据核字 (2015) 第 175281 号

姚仁喜｜大元建筑作品 30x30

著　　者：姚仁喜
编辑总监：刘玉贞　姚任祥
编辑执行：温淑宜　乔苹　林宜熹
美术总监：段世瑜
美术编辑：方雅铃　陈怡茜　郑乃文
摄　　影：张全筴　郑锦铭　陈弘昑　潘瑞琮　黎不修　游宏祥　马怀仁　Marc Gerritsen
　　　　　Bruno Klomfar　Willy Berré　THSRC
策　　划：彭礼孝　柳青

出版发行：辽宁科学技术出版社
　　　　　（地址：沈阳市和平区十一纬路29号 邮编：110003）
印　刷　者：北京雅昌艺术印刷有限公司
幅面尺寸：300mm×300mm
印　　张：16.5
字　　数：20千字
出版时间：2015年8月第1版
印刷时间：2015年8月第1次印刷
责任编辑：包伸明　张翔宇
责任校对：王玉宝

书　　号：ISBN 978-7-5381-9372-5
定　　价：160.00元